Internet of Things for Smart Buildings

Leverage IoT for smarter insights for buildings in the new and built environments

Harry G. Smeenk

‹packt›

BIRMINGHAM—MUMBAI

Internet of Things for Smart Buildings

Group Product Manager: Mohd Riyan Khan

Publishing Product Manager: Suwarna Rajput

Senior Editor: Arun Nadar

Technical Editor: Shruthi Shetty

Copy Editor: Safis Editing

Project Coordinator: Ashwin Kharwa

Proofreader: Safis Editing

Indexer: Hemangini Bari

Production Designer: Vijay Kamble

Marketing Coordinator: Nimisha Dua

First published: March 2023

Production reference: 1100323

Published by Packt Publishing Ltd.

Livery Place

35 Livery Street

Birmingham

B3 2PB, UK.

ISBN 978-1-80461-986-5

www.packtpub.com

To my loving and supportive wife of over forty years, Lynn, for listening to my nonsense about IoT, smart buildings, and music. To my children, Nicole and Matthew, my son-in-law Adam, and our incredibly awesome grandchildren, Bradley and Norah; Bradley keeps me on my technology toes. To the memory of my loving mother and to the memory of my father-in-law, James Phillips, who was always there for us.

Foreword

Smart buildings are not new. They began in the 1980s and continue to be a major topic of discussion throughout the built environment. However, today, the quest for smart buildings is in its most transformative and compelling period we have seen over the last twenty-five years, driven by several influences, of which the **Internet of Things (IoT)** is one.

With the introduction of IoT, we have seen tremendous advancement occur with technologies and solutions that have enabled the convergence of information technology, operational technology, and building automation. The adaption of building system integration and IoT building technologies continues to expand beyond the commercial office building, spreading across to hospitals, hotels, retail, schools, universities, and manufacturing. Also, the next-generation IoT solutions have opened new advancements in how we manage and operate facilities.

Furthermore, IoT has caused a shift in the value equation; we are moving beyond efficiency to a more holistic view encompassing the overall performance of buildings and their equipment systems as assets that are being utilized for increasing value. This is driving increased collaboration across business functions and the entire building ecosystem. And it is not just the economic factors that are being captured with simple ROI calculations; it is a combination of the economics and the rising expectations of building owners and operating managers who increasingly live in a technology environment that is more advanced than their building systems.

IoT devices that connect directly to the enterprise are accelerating. These new devices are smarter, more powerful, and offer higher levels of functionality with enhanced embedded systems software. The trend is toward connecting more devices that provide information within each device. The additional value will take shape when the devices are extended by layering applications that leverage the activities, services, and interrelationships—not only of the devices but of all people, systems, and connected devices in the network.

Smarter buildings are now about providing what operators need to do and whatever it is they wish to do. We need to get better at delivering solutions that solve owner/operator challenges (problems) and achieve their business and strategic outcomes.

Transformation and innovation via IoT will continue driving smart buildings.

In Harry G. Smeenk's book, *Internet of Things for Smart Buildings*, he addresses a range of topics based on his 40 years of delivering and implementing networks in all types of buildings and his work on developing smart building assessments and certifications. Topics include architecture, controls and automation, the smart building stack, energy management, preventive/predictive maintenance, data, cybersecurity, and systems such as **heating, ventilation, and air conditioning** (**HVAC**), lighting, occupancy **indoor air quality** (**IAQ**), space management, and safety.

Whether you are a building owner, operator, facility management, system integrator, contractor, engineer, or service provider, want to deliver your first smart building, or you have delivered several, you will find the level of detail exceptional. This book is an invaluable resource for those who want to make the most of creating and delivering smart buildings.

The adoption of smart building technology is at an inflection point driven by IoT. The industry is currently being influenced by several forces creating this inflection point. The influence of business outcomes, flexible workplaces, demand for IAQ, the increasing number of connected devices and systems, cyber threats, and the global drive toward net zero all depend upon connected and smarter buildings.

Marc Petock

Chief Marketing and Communications Officer, Lynxspring

Contributors

About the author

Harry G. Smeenk is a technology strategist and thought leader in smart buildings, IoT, edge computing, and networks. He is an executive leader in the design, development, deployment, and integration of smart building IoT networks with Tapa Inc and Smart Buildings Online LLC. He has driven worldwide technology roadmaps, best practices, and standards for the Telecom Industry Association. He conceptualized and developed the industry's first smart building rating program. As an entrepreneur-in-residence at the North Texas Enterprise Center, he helped launch and accelerate start-ups, including three of his own. He earned an MBA from the University of North Carolina and a bachelor's in business management from St. John Fisher College.

About the reviewers

Anirudh Bhaskaran is an accomplished professional with extensive experience in business strategy and market intelligence. He is currently associated with Frost & Sullivan, a leading global management consulting firm, as an industry principal in energy and environment practice specializing in the buildings and smart infrastructure industry. As a key member of the organization, he leverages his expertise and experience to provide clients with valuable insights and recommendations, helping them make informed business decisions. Anirudh Bhaskaran received a bachelor's in mechanical engineering from Rajalakshmi Engineering College and a master's in energy engineering from PSG College of Technology.

I would like to thank my whole family for their unwavering support and love, which has been a constant source of strength and comfort for me throughout my life. I am grateful to them for tolerating my busy schedule, as it takes a lot of time to research and validate data and insights in the field of market research and consulting.

Tulshiram Waghmare is an experienced technology and industry professional and a proponent of smart and healthy building technology. He is a successful professional in developing and managing products based on IoT, with over 20 years of experience.

He is passionate about solving the global warming issue and strongly believes that smart buildings driven by technology have immense potential to reduce **greenhouse gas** (**GHG**) emissions and operating costs and enhance occupant health and productivity. Tulshiram has extensive experience in building management systems, energy performance contracting, and industrial automation besides his current role in digital product development for smart and healthy buildings.

Thanks to everyone who worked on the book in the Packt team who helped me.

Table of Contents

3

First Responders and Building Safety · 43

4

How to Make Buildings Smarter with Smart Location · 61

5

Tenant Services and Smart Building Amenities · 75

Part 2: Smart Building Architecture

6

The Smart Building Ecosystem   93

7

Smart Building Architecture and Use Cases   111

Part 3: Building Your Smart Building Stack

11

Technology and Applications   179

Part 4: Building Sustainability for Contribution to Smart Cities

12

A Roadmap to Your Smart Building Will Require Partners   199

Preface

Buildings are a **System of Systems** (**SoS**), each controlling a vital function to achieve the building's purpose. Smart buildings use **Internet of Things** (**IoT**) devices, sensors, and software to monitor and control various building functions to optimize the building's environment and operations. As a result, they improve building efficiency, reduce costs, reduce resource usage, and improve occupant satisfaction.

Almost every function within a building is a candidate for building your smart building applications with IoT devices. Whether you start with one function or multiple functions, this book will review the many opportunities and technologies used to build a smart building. Edge routers, numerous IoT sensors and devices, the various connection options available, and the software required both locally and via the cloud to make it all work together seamlessly are discussed.

This book will leverage my smart building experience to help you to do the following:

- Make a building's mechanical, electric, HVAC, and networks smart with the addition of IoT, connectivity, and software solutions
- Get multiple building systems (HVAC, security, fire, and so on) to communicate with each other to analyze data and deliver actionable operations outcomes
- Improve operational efficiency, reduce waste and carbon footprint, consume fewer resources, improve productivity, and enhance the occupant's quality of experience
- Understand the communications protocols, methodologies, components, and architecture used to deliver IoT for smart buildings
- Solve key building industry issues such as return-to-building, proprietary software languages that prevent other vendors from connecting, lack of IoT and smart buildings skills, and meeting government energy and environmental requirements

It is the only comprehensive book explaining nearly every aspect of how IoT makes buildings smarter. You will learn about terminology, technology, standards, methodologies, and frameworks to help you decide which projects are right to build your smart buildings.

It will provide application examples, definitions, and detailed explanations of what building systems can be made smart with IoT. Various technologies and architectures are reviewed to demonstrate how to design and implement solutions, including how to use IoT stacks. By the end of the book, you'll be able to identify and design your own smart building initiatives to solve building-related challenges.

Who this book is for

This book is for architects; mechanical, electrical, and HVAC engineers; system integrators; facilities and operations personnel; and others looking to implement IoT solutions to make their buildings smart.

This book reveals examples, strategies, and frameworks that will allow your business to realize sustainable financial benefits and efficiencies from smart buildings.

The reader should have a basic knowledge of the various mechanical and electrical building systems, including HVAC, security, fire alarms, communications, and data networks, as well as the operations and maintenance requirements.

What this book covers

Chapter 1, An Introduction to IoT and Smart Buildings, lays the foundation by defining the IoT and smart buildings and how they interact. Traditional building issues will be identified, and we'll discuss how IoT smart building solutions resolve these. It also covers the history and evolution of building control systems and how smart buildings contribute to smart cities.

Chapter 2, Smart Building Operations and Controls, explains how a building comprises several systems, each operating independently with a few connected to each other. This chapter shows how these systems benefit enormously by adding IoT devices that can monitor, measure, report, control, and optimize various functions when connected to a common IoT network.

Chapter 3, First Responders and Building Safety, demonstrates how IoT helps first responders to better understand the building with access to a visual display of the building's footprint, which draws on both real-time situation alerts forwarded from dispatch as well as stored building floor plan and firefighting equipment location data. You'll learn how the incident commander can see the building footprint and interior plans, the location of fire hydrants, and how building operations can achieve significant improvements in building access, security, and communications.

Chapter 4, How to Make Buildings Smarter with Smart Location, identifies how building data can be enriched with the location context. This chapter demonstrates how location-based services are being used to increase efficiency, improve safety, and enable a more enhanced user experience. It will demonstrate that smart locations are physical locations equipped with networked sensors to give owners, occupants, and managers more information about the condition of those locations and how they're used.

Chapter 5, Tenant Services and Smart Building Amenities, outlines how building owners and operators introduce IoT and smart building solutions to improve their operational efficiency (therefore reducing cost) and to improve occupant satisfaction (that hopefully increases revenues). This chapter will explore the numerous applications that have been developed to improve almost every aspect of building management and occupant interaction points.

Chapter 6, The Smart Building Ecosystem, shows how new buildings incorporate IoT solutions as part of their initial design, and existing buildings add IoT components to make the building smarter. This chapter highlights the five major components; IoT sensors and devices, edge or cloud computing, analytics software, a user interface, and a means of connectivity to produce enormous amounts of data to manage a smart building.

Chapter 7, Smart Building Architecture and Use Cases, demonstrates how smart buildings today are not designed from the top down; rather, they are assembled from the bottom up, pulling together components that are independently designed and implemented separately from each other. This chapter reviews the various components, the importance, and the challenges of developing a smart building architecture and review several use cases. NIST cybersecurity best practices are introduced.

Chapter 8, Digital Twins – a Virtual Representation, demonstrates the digital replica of physical assets, processes, people, places, systems, and devices used for various purposes within the building. It shows how a digital twin virtual representation of the physical building is embedded with rich information about spaces and assets that can offer significant benefits to building owners.

Chapter 9, Smart Building IoT Stacks and Requirements, discusses how the complexity of smart buildings can be overwhelming, especially when there are numerous vendors, products, and technologies involved. This chapter introduces the Smarter Stack used to map existing building systems, IoT devices, and technologies to compare or identify gaps in a vendor's product or from a customer's requirement perspective.

Chapter 10, Understanding Your Building's Existing Smart Level and Systems, demonstrates that to begin any smart building undertaking, you must first understand what the current systems are and how they are configured and connected, and then determine what modifications and new systems will be required. This chapter introduces several industry smart building assessment programs to determine the current level of a building's smartness.

Chapter 11, Technology and Applications, focuses on the technology and applications required to make the building smart. It examines an extensive list of smart building application opportunities available to deliver smart requirements, along with examples for each. It defines the role of middleware in delivering these applications and concludes with a review of the codes, standards, and guidelines to be considered to prevent becoming locked into proprietary solutions that may prevent expansion later.

Chapter 12, A Roadmap to Your Smart Building Will Require Partners, offers a roadmap for existing and new buildings to make your building smarter as each new system is integrated. It makes references to previous chapters to indicate where that chapter's subject matter fits into the roadmap and identifies the various partners that may be required beyond the construction crews.

Chapter 13, The Importance of Smart Buildings for Sustainability and the Environment, reminds us that smart buildings use IoT to share information, control operations, and enhance human interaction. In addition, because buildings require a lot of energy to operate, smart buildings are equipped to better manage energy usage, and this chapter provides ways to reduce carbon footprint, foster sustainability, and endorse eco-friendly alternatives.

Chapter 14, Smart Buildings Lead to Smart Cities, theorizes that buildings are an ideal starting point from which to grow smart cities. This chapter demonstrates how buildings are a microcosm of a city with similar needs to manage resources, water, energy, lighting, emergency services, security, and other services. Along the same line of reasoning, smart buildings are a microcosm of smart cities and therefore serve as the ideal launching point to grow and develop smart cities.

Chapter 15, Smart Buildings on the Bleeding Edge, discusses how the growing development of cloud computing and data management links together multiple data sources, inputs, and user types into a cloud of useful information to create a more efficient, effective, and engaging smart building. But what's next? This chapter will explore the evolution of smart buildings, introduce the unified building, and list what many consider to be the smartest buildings and cities at the end of 2022.

To get the most out of this book

The reader should have basic knowledge of the various mechanical and electrical building systems, including HVAC, security, fire alarms, communications, and data networks, as well as the operations and maintenance requirements.

Download the color images

We also provide a PDF file that has color images of the screenshots and diagrams used in this book. You can download it here: `https://packt.link/Q6g3b`.

Get in touch

Feedback from our readers is always welcome.

General feedback: If you have questions about any aspect of this book, email us at `customercare@packtpub.com` and mention the book title in the subject of your message.

Errata: Although we have taken every care to ensure the accuracy of our content, mistakes do happen. If you have found a mistake in this book, we would be grateful if you would report this to us. Please visit `www.packtpub.com/support/errata` and fill in the form.

Piracy: If you come across any illegal copies of our works in any form on the internet, we would be grateful if you would provide us with the location address or website name. Please contact us at `copyright@packt.com` with a link to the material.

If you are interested in becoming an author: If there is a topic that you have expertise in and you are interested in either writing or contributing to a book, please visit `authors.packtpub.com`.

Share Your Thoughts

Once you've read *Internet of Things for Smart Buildings*, we'd love to hear your thoughts! Scan the QR code below to go straight to the Amazon review page for this book and share your feedback.

https://packt.link/r/1804619868

Your review is important to us and the tech community and will help us make sure we're delivering excellent quality content.

Download a free PDF copy of this book

Thanks for purchasing this book!

Do you like to read on the go but are unable to carry your print books everywhere?

Is your eBook purchase not compatible with the device of your choice?

Don't worry, now with every Packt book you get a DRM-free PDF version of that book at no cost.

Read anywhere, any place, on any device. Search, copy, and paste code from your favorite technical books directly into your application.

The perks don't stop there, you can get exclusive access to discounts, newsletters, and great free content in your inbox daily

Follow these simple steps to get the benefits:

1.   Scan the QR code or visit the link below

https://packt.link/free-ebook/9781804619865

2.   Submit your proof of purchase

3.   That's it! We'll send your free PDF and other benefits to your email directly

Part 1: Applications for Smart Buildings

Smart buildings use the **Internet of Things (IoT)** devices, sensors, and software to measure, monitor, and control various building characteristics to optimize the building's operations and environment. Part 1 examines the key building systems that may benefit from *smart* applications.

This part contains the following chapters:

- *Chapter 1, An Introduction to IoT and Smart Buildings*
- *Chapter 2, Smart Building Operations and Controls*
- *Chapter 3, First Responders and Building Safety*
- *Chapter 4, How to Make Buildings Smarter with Smart Location*
- *Chapter 5, Tenant Services and Smart Building Amenities*

1

An Introduction to IoT and Smart Buildings

Smart buildings use **Internet of Things (IoT)** devices, sensors, and software to monitor and control various building functions to optimize the building's environment and operations. They improve building efficiencies, lower costs, and improve occupant satisfaction. What constitutes a smart building? How many IoT sensors are required, and how many applications are needed to determine the degree of building smartness?

Internet of Things for Smart Buildings is a comprehensive guide for those who want to build either a greenfield smart building (new) or retrofit a built environment. Almost every function within a building is now a candidate for building smart building applications and IoT devices. Whether you start with one function or multiple functions, this book will review the many opportunities and technologies used to build a smart building. Edge routers, numerous IoT sensors and devices, the various connection options available, and the software required both locally and via the cloud to make it all work together seamlessly will be discussed.

Due to the various technologies, and the number of vendors and products involved, smart building projects can be complex and sometimes overwhelming. **Smart building stacks** can be used to map building products, IoT devices, and technologies for comparison or to identify gaps in a vendor's product or from a customer's requirement perspective. Complete with product, solution, and technology descriptions, examples, and recommendations, this book will help you decide which projects are right to build your smart building, and show you how to develop your technology and business stack.

This chapter will lay out the basics by defining what a smart building is and the contribution of the IoT to smart buildings. We'll start by giving an example of a person's typical day working in a smart building and the many smart features and benefits they may come across. We'll review traditional building issues and the benefits of developing a smart building solution to resolve these issues.

To better understand how the industry got here, we will provide a basic history and evolution of building control systems, intelligent buildings, and the IoT. We will discuss the many benefits of a smart building for building owners, operators, occupants, and the community. Finally, we will introduce how smart buildings contribute to smart cities.

In this chapter, we're going to cover the following main topics:

- What a smart building is with an example of a day in a smart building
- What the IoT is
- How smart buildings and IoT came together
- The benefits of having a smart building
- A review of traditional building issues and how smart buildings solve these
- Why we need smart buildings to have a smart city
- The history of building control systems and evolution to smart buildings

A day in a smart building

Imagine for a moment that you are a highly experienced building engineer working for a property management company in their regional office located in an eight-floor downtown commercial office building. Imagine this building is a smart building and what a typical day might look like. This section will provide an example of the many smart building benefits and features this person might encounter. The example is used to place you in the smart building right away to begin to understand what a smart building is from its numerous applications and functions.

Arrival and access

You arrive at the building's parking lot in your **Electric Vehicle** (**EV**), and you are notified via your smartphone or car display that spot #304 on the third floor is the closest EV charging spot available. Almost 40% of parking spots are EV charging spots in this smart building versus 2-5% in non-smart buildings. As you approach the building, your smartphone displays the real-time **Indoor Air Quality** (**IAQ**) score for the entrance lobby, the elevator banks and elevators, your floor's lobby, and your office.

You are comforted to know that this smart building uses the very latest in IoT environmental sensors to continuously measure the air quality in your office, conference rooms, and common areas such as breakrooms and restrooms. In a post-pandemic environment, indoor air quality standards, regulations, and recommendations have been implemented around the world to ensure building occupants have safe air quality levels. An example of how IAQ constantly changes during the day is in a conference room. As more people enter the room, carbon dioxide (CO_2) levels may increase beyond recommended levels, requiring the building's **Heating, Ventilation, and Air Conditioning** (**HVAC**) system to open air dampers to let more fresh air in to keep the IAQ score within recommended ranges.

The state-of-the-art **Building Management System** (**BMS**) uses IoT sensors and controllers to monitor and control all the building's various systems. These systems typically include HVAC, safety and security, cameras/CCTV, fire alarms, lighting, water, gas, other fluids, networking, communication, and so on. These systems provide a wealth of information about how the building is performing and are all well-suited for automation for efficiency improvements and cost reductions.

As you enter the building, occupancy sensors, facial recognition cameras, card access readers, smartphone QR codes, or any other IoT device solution approves your entrance, opens the doors, and registers you as *entered the building*, either as an occupant or guest. Guests may pre-register their information prior to arrival or they may be directed to a reception desk, a kiosk, or a phone system to complete registration. For existing occupants entering the building, a wealth of stakeholder-approved information is now accessed indicating the occupants' preferences.

Information, preferences, and data

Using the occupant's stored information, the smart elevator flashes a personalized welcome screen, selects an elevator car, opens the door automatically, and delivers the occupant to the correct floor, all touch free for added safety. Digital information is provided in the elevator highlighting that car's IAQ, today's building events and announcements, along with today's cafeteria lunch special. As the elevator is on its way, your office lights turn on, the temperature is automatically adjusted to your preference, and your favorite music begins to play.

Once in your office, you immediately begin checking the numerous systems and portfolio buildings your company monitors in your **Virtual Network Operations Center** (**VNOC**). This requires enormous amounts of bandwidth, which is provided in this smart building via a high-bandwidth fiber-optic cable backhaul network. Using fiber-optic cables provides for greater capacity, increased speeds, reduced latency, and better supports a number of different technologies. A drone delivers your coffee directly to your desk and a FedEx in-building robot delivers your documents. Another robot fills components and supplies for inventory in the storage area.

As you prepare for the team meeting scheduled later in the day, you take a few minutes to review the real-time and historical energy management information for your office and the building. Recently, IoT sensors and controllers were installed in the energy management system along with an **Artificial Intelligence** (**AI**) platform to deliver energy reductions of 20% to 30%. To monitor and improve energy efficiency, an energy accounting system measures, analyzes, and reports on the energy consumption of different activities on a continuous basis.

As part of the smart building's lighting system, the daylight harvesting system can reduce energy consumption by using daylight to offset the amount of electric lighting needed by dimming or brightening the lights based on the actual daylight available. LED bulbs, voice activation, and motion sensors, along with lighting system management software, allow for additional scheduling and control capabilities.

Since many government agencies across the world offer tax deductions, funding, and/or other incentives for the costs of improving the energy efficiency of buildings, you begin to check to see what programs are available for your building. You note there is a deduction available for the entire cost of installing energy-efficient systems and smart meters in your building and you notify the accounting department to claim these credits. You also note that last month, your building's solar panels were able to give back to the smart city grid and you'll need to copy that data report to claim the incentives that are available via the **Grid-interactive Efficient Buildings** (**GEB**) initiative in your area. There may also be carbon footprint credits available you'll want to check on.

Communications network and collaborative conferencing

As your smartphone rings, you are still amazed by the incredible in-building coverage you have for all your devices through the building's secure private network. The building owner decided to treat communication and networking as a fourth utility in the building by delivering a highly secure system that provides coverage to 99.9% of the building, including elevators, stairwells, basements, and the parking garage. This privately owned system is comprised of vertical and horizontal fiber runs, coupled with 5G cellular small cell and **Distributed Antenna Systems** (**DASs**) as part of the **Radio Access Network** (**RAN**). Wi-Fi, point-to-point commercial microwave links, and Bluetooth are also used to connect many IoT devices. Fiber-optic backhaul delivers high bandwidth and reduces latency for overall control.

It's now 10:00 and you have a meeting scheduled in one of the many smart conference rooms. Since you are now working on a hybrid model, some of your team will be remote today and the smart conference tools will keep the meeting moving forward. You scheduled the room last week using a tenant application provided by the building manager. Lighting levels and the temperature have been automatically set to your profile preferences and the delivery robot has brought drinks and water for everyone.

With collaborative conferencing technology and AI tools, your presentation is automatically loaded from your computer, and audio and visual systems have been automatically turned on. The conference call autodial has launched, and team members are ready to begin the meeting right on time. Wireless broadband and airplay allow others to share data in real time during the meeting. The smart jam whiteboard captures all meeting information and notes, and the auto transcription system automatically captures and develops meeting notes that are sent directly to each participant immediately at the meeting's conclusion.

During the meeting, you are asked to explain the smart privacy windows installed throughout the building. These sustainable, energy-efficient windows also act as transparent solar panels. Some may even contain antennas to help bring the cellular network into the building. Each window has a unique **Internet Protocol** (**IP**) address used to identify itself and communicate with other devices. Since they tint automatically or via remote control, no bulky expensive blinds are required, and they control heat and glare. Immersive display windows/glass may also transform into digital, interactive surfaces. The

View smart window company suggests workplace productivity and wellness are improved when such windows are in place, through improved moods, reduced eyestrain, fewer headaches, and less drowsiness.

The meeting is running long so we check to see that the conference room is available for another 2 hours, and we decide to work through lunch. Each person orders lunch from the building's cafeteria via their smartphone and an autonomous delivery robot arrives 30 minutes later with everyone's orders. These delivery robots are helping the building's owner bridge the gap arising from recent labor shortages.

You are alerted via your smartphone that your guest speaker has arrived in the building. As she is registering with the front desk, a wayfinding application is automatically delivered to her smartphone. Wayfinding will help her navigate her way around the building and other unfamiliar environments. It safely manages the movement and flow of people through the building, while encouraging social distancing. It improves the user experience and contributes to a sense of well-being and security. It also saves time as you are not required to leave your meeting to go down to the front entrance to escort your guest.

Smart building infrastructure

This guest is a power vendor, and she is there to deliver a presentation on unique low-voltage power systems now being considered for the building. A few different alternatives are being explored, starting with **Power over Ethernet** (**PoE**). In a POE system, electric power and data are transferred using Ethernet cables. Savings are achieved by eliminating separate power supply cabling and outlets. **Power over Fiber** (**PoF**) is similar, whereby the fiber-optic cable carries optical power to supply the energy source, and data is transmitted over the same fiber cable.

The meeting will conclude with two short education and training sessions. The first session is on meeting regulatory requirements for your building. There are specific national, provincial, state, county, city, town, and local jurisdiction requirements that vary greatly across countries, regions, and continents. Our instructor does not want to give another boring slide presentation, so they hand us all **Virtual Reality** (**VR**) headsets. VR training delivers on-the-job training using real-life settings through an immersive learning experience. By learning by doing, skill retention is high, and workplace productivity is increased.

The second presentation is on new tools that are being used to design and construct or retrofit buildings. **Building Information Modeling** (**BIM**) allows us to create a digital twin record of our facility information such as blueprints, emergency plans, plumbing, and electrical installations to store them digitally. 3D laser scanning or building thermal imaging surveys may save time, reduce rework, restore missing data and drawings, reduce liability, and minimize risk. Using these, one can scan any type of building and receive a **Computer-Aided Design** (**CAD**) rendering. 360-degree panoramic images are also available. These can be used to reverse engineer an existing building. Detailed mapping is achieved in every location within a building using the latest 3D scanning techniques, drones, **Unmanned Aerial Vehicles** (**UAVs**), terrestrial scanners, and digital photogrammetry.

Later, you are notified via a text message that the predictive maintenance system has identified that a part will need to be replaced based on the mean time between failure calculations. **Mean Time Between Failures** (**MTBF**) uses historical data to calculate the average time between component and system breakdowns. MTBF is considered a critical tool in the maintenance program to measure performance, safety, and equipment design. The text notification indicated that the parts were ordered and the auto-scheduler was scheduling the installation time and maintenance engineer.

Predictive maintenance technologies are designed to identify potential maintenance issues before they become a problem. With the rise of smart sensors and IoT, these sensors make maintenance smarter, cheaper, and more efficient. These sensors are installed on or near radiators, boilers, pumps, and other machinery. They detect critical levels of noise, vibration levels, leaks, or changes in temperature, and when a certain threshold is achieved, our smart system automatically orders the part and schedules the repair before the issue escalates into a system failure.

Our Smart Building Asset Management program tracks, manages, monitors, and plans our IT assets for all our buildings. We collect the correct information associated with each asset and then assign and collect asset information by property to manage, track, evaluate, and assign costs appropriately. Vendors are assigned to the assets to capture support detail information for easy and efficient vendor support. Cost information is assigned for cost tracking and allocation and to assist us in making information-based decisions regarding our building's assets. IoT sensors are placed in or on each asset to provide real-time positioning, location, and tracking.

Tenant applications

Your final appointment for the day involves onboarding your newest tenant to the building. Each tenant is provided with our smart building **Tenant Mobile Application** (**TMA**), which is designed to give the tenant more control over their environment. The app also will digitalize our tenant-facing services, transactions, and work orders, to enhance the overall **Quality of Experience** (**QoE**) for the building's occupants with numerous amenity services.

From this smart building app, your building occupants can manage tenant services such as submitting and tracking their tenant request work orders, along with receiving cost options and estimates if required. Tenants directly control temperature and lighting along with other comfort settings, eliminating costly building engineer work orders for these. They can check space availability and make reservations for conference rooms, desks, and common-use areas. They can see real-time space occupancy and trend space utilization rates. They may pay their lease rent and fees via the app, and they have access to historical payment information. Incident management and insurance information are also part of the tenant services app.

The smart building app delivers real-time access to the building's amenities and concierge services. Parking, EV charger availability, bike parking, and storage locker information are available. Using occupancy sensors, the app provides real-time access to the wellness center, yoga studio, and workout facility information to assist tenants in scheduling their workouts. Information regarding public transportation, ride-share services, taxis, and limo services for the building provide direct app access along with pick-up and drop-off location information with IoT cameras for real-time checks.

Food, beverage, and entertainment access coupled with any in-building retail and other amenity marketplace access and information are included along with the building and local community event calendars. Customized apps have been developed for many of our tenants with documents such as human resource manuals, safety manuals, training guides, and surveys. A gamification app was even developed for a high-tech tenant.

In your residential buildings, your renters prefer smart apartments, which make their lives easier, more convenient, and more fun. These smart apartment amenities help tenants automate routine tasks, save money, and to find time to do other things they prefer to be doing. A long list of smart apartment amenities includes smart locks, smart thermostats, smart lighting, a single residential mobile app, Wi-Fi-as-a-service, smart access control, voice control, smart security systems, instant messaging, community events, ridesharing pick-up/drop-off locations, automated package management, in-unit package drop-off, bike sharing, automated maintenance requests, smart appliances, and instant payments. Smart apartment amenities are gaining popularity and there is no end in sight.

In a post-pandemic environment, commercial office buildings, factories, hospitals, and many other building types are making adaptations that work well for social distancing needs. Wide-open floor plans and the ability to scale, move, and function with less physical touch are the new normal. Your smart building includes smart access systems and smart elevators as mentioned earlier. Your updated restrooms include IoT touch-free sinks, paper towel dispensers, and toilets, which also alert building maintenance when soap, toilet paper, and paper towels are running low. Your HVAC filters have been upgraded to hospital-grade ones with built-in IoT sensors that notify you when they need to be replaced. Air and surface disinfectants are designed to mitigate germs and are monitored with IoT sensors.

Your day in the smart building is coming to its conclusion. Your electric car is fully charged, and you have ordered a take-out dinner from the restaurant on the first floor. You are waiting for the delivery robot to meet you in the lobby with your dry cleaning and you have called the elevator from your smart app. As you leave the office, the lights and music are turned off, and the temperature is adjusted to help lower your energy costs.

Hopefully, this hypothetical example of a day in a smart building has begun to highlight the many different aspects and components that can be combined to build your smart building. There is no magical number of components, applications, or amount of data that is collected that constitutes crossing over into the *officially a smart building* category, but recent industry collaborations are beginning to provide assessment criteria to determine just how smart your building might be and what steps might be required to move to the next level.

Another highlight from your day in the smart building example is that no two buildings will be the same (unless intentionally built as duplicates). While it is obviously easier to build a smart building with a new construction, cost considerations may value-engineer many desired features out. Since most buildings are in the built environment, retrofits and upgrades that contain IoT devices and sensors will be the typical method for transitioning to a smart building. With one target area IoT project or many IoT projects, you'll be on your way to building your smart building.

Smart building definition

This leads us to the definition of a smart building. Ask many people and you will receive many answers, as there is not a clear definition:

- According to Paul Wellener et al., at Deloitte Insights, "*Smart buildings are digitally connected structures that combine optimized building and operational automation with intelligent space management to enhance the user experience, increase productivity, reduce costs, and mitigate physical and cybersecurity risks.*"

- The **Telecommunications Industry Association** (**TIA**) defines a smart building as "*all building systems are fully integrated and sharing data, so they be managed through a single pane of glass with minimal human intervention.*"

- Ernst & Young Global Ltd. (EY Group) states "*a smart building can be thought of as an ecosystem, a dynamic entity with many devices of varying age that 'talk to' and depend on one another, sharing data and responding to various needs. Key to this complex interaction of software and hardware is the human element, the overriding "voice," if you will, of guidance and direction that points all other systems toward those goals*".

Throughout this book, we will use the following definition: "*A smart building uses an integrated set of technology, systems, and infrastructure to optimize building performance and occupant experience*".

Types of smart buildings

Throughout this book, we will refer to smart buildings, meaning any type of building. Each building type can implement some or all of the smart solutions we cover throughout the book; however, in this list, we have indicated some focused applications for building types based on their primary function.

Building Category	Building Examples	Focused Smart Building Applications
Agricultural buildings	Barns, greenhouses, silos, coops	Temperature, humidity, lighting
Commercial buildings	Multistory with at least 50% used for commercial activities such as restaurants, retail, shops	HVAC, energy, occupancy, asset tracking

Building Category	Building Examples	Focused Smart Building Applications
Data centers	Standalone and mixed-use data centers	Temperature, humidity, security
Education	Universities, schools, colleges, daycare, technical	Air quality, occupancy, access, security
Event buildings	Stadiums, arenas, theaters, auditoriums, conference centers,	Occupancy, safety, security systems
Government or civic buildings	Courts, post offices, tax offices, jails, admin buildings, museums, police and fire stations, military, community centers, libraries, and so on	Air quality, occupancy, access controls
Hospitality	Lodging, hotels, motels, resorts, historic inns, boutique hotels, B&Bs, cruise ships	Guest amenities, cleaning, security, energy
Industrial or manufacturing	Buildings for manufacturing, production, assembly, repairs, altering, renovating, ornamenting, power plants, water plants, and so on	Asset tracking, wayfinding, spills, leaks, preventative maintenance
Medical	Hospitals, medical offices, local ER shops, doctors' offices, clinics	Navigation, patient tracking, wayfinding
Office buildings	One-story, multistory, campuses, mostly used for offices	HVAC, energy, air-quality systems, workflow
Owner occupied	Typically, a company owns and uses the building for its company's needs	Employee workflow tools. Air quality
Residential / **Multi-Dwelling Units (MDUs)**	Apartments, condominiums, townhomes, dormitories, MDUs, nursing homes	Access, security, safety systems, air quality
Recreational buildings	Fitness centers, bowling alleys, gyms, ice rinks, indoor swimming pools	Access, security, safety, cleaning
Religious	Churches, temples, synagogues, temples, mosques, cathedrals, monasteries	Access, security, safety, air quality
Retail	Stores, malls, shops, big-box stores, grocery stores	Occupancy, security, asset tracking
Transportation	Airports, train stations, bus terminals, subway stations, ferry stations, others	Occupancy, security, safety systems
Warehouses	Private, public, climate-controlled, distribution centers, storage	Asset tracking, wayfinding

Table 1.1 – Types of smart buildings and their applications

What is the Internet of Things (IoT)?

IoT connects and exchanges data from physical objects to other physical objects using processing ability, software, and other technologies. While the term *internet* is used, it is considered a misnomer since devices do not need to be connected to the public internet. In many applications today, devices and sensors are connected to a private network where they can be individually addressable or even connected directly to each other.

IoT crosses many industries and markets and is not limited to buildings; however, this book's focus will be on its application in buildings. Most people today learned of IoT through the consumer market and through smart homes with features such as smart lighting, thermostats, cameras, security systems, and smart appliances. Other common applications today include smart speakers, smart watches, and healthcare devices – all IoT connected to our smartphones.

Numerous industrial applications use IoT devices to collect and analyze data for connected equipment and often are referred to as **Industrial Internet of Things (IIoT)**. **Operational Technology (OT)** is often combined with IIoT to regulate and monitor industrial systems and manage assets. Other IoT applications include manufacturing, agriculture, energy, environmental, military, and metropolitan systems to manage cities and utilities.

While many people believe that IoT is a recent technology development, the concept of smart devices was introduced in 1982 with a modified Coca-Cola machine becoming the first connected device. In September 1985, Peter T. Lewis introduced the concept and term *Internet of Things* for the first time to the Congressional Black Caucus Foundation's 15th Annual Legislative Weekend in Washington, D.C. It wasn't officially named the Internet of Things until 1999 by MIT's Executive Director of Auto-ID Labs, Kevin Ashton. During his presentation to Procter & Gamble, he described IoT for the first time and the definition has evolved since then.

IoT challenges

IoT devices are used to monitor and control many of a building's electrical, mechanical, and energy management systems to improve efficiencies and reduce costs. **Building IoT (BIoT)** suffers from platform fragmentation, interoperability issues, and lack of standards making it sometimes difficult for devices to speak to each other. Coupled with the numerous options for connecting these devices with fiber-optic cables, copper cables, and an endless list of wireless connectivity options, IoT use in smart buildings creates some challenges that must be considered and managed.

With the large amounts of surveillance sensors and the data that is collected and stored, privacy threats are enormous, as is the potential for hackers to create disruption and misuse the information. Many people are concerned that companies and governments collecting this data are also selling it, making us more transparent and making it harder for us to control our privacy.

Data storage challenges include how and where to store all this data, either locally, in the cloud, or in a data warehouse. Questions about how long data should be stored are usually answered by legal requirements and the cost of storage. A return-on-investment business analysis usually answers how long the data will be stored because it is expensive to store data. Access to this data raises other concerns as to who should be able to see and use the data. Another IoT data challenge is how to tag it for commonality and for easier reference later. Today, there are numerous tagging languages being used that are not interchangeable.

Security is one of the biggest challenges and concerns with IoT and smart buildings. These include authentication concerns, unencrypted messages sent between devices, poor handling of security updates, man-in-the-middle attacks, and breaches. It is important to understand that if everything is connected, then access is also connected. A recent massive data breach at retail giant Target allegedly resulted partly from their failure to properly segregate systems and payment card data. According to Jaikular Vijayan's February 6, 2014, *Computerworld* article *Target breach happened because of a basic network segmentation error*, the payment data was stolen by hackers using stolen login credentials for the HVAC system and then moved about undetected on Target's network. IoT systems control large amounts of safety system sensors such as smoke detectors, contact sensors, motion sensors, door access controllers, and numerous others. Potential challenges include device or communication failures, software bugs, or other unforeseen bad app interactions, all of which could cause an unsafe or dangerous physical state.

Design challenges include solving communications issues between various systems, confusing terminology, scalability, environmental/sustainability impacts, device obsolescence, and lack of interoperability. Additionally, designs need to consider organization capabilities, cultural requirements, industry standards, and the numerous governmental codes, regulations, and laws from the various agencies and departments.

Finally, a mere attempt at a smart building IoT project creates unique business planning and project management challenges. IoT projects run differently than simple, traditional IT, manufacturing, or construction projects. Smart IoT projects are complex, and designers and project managers are generally inexperienced in this area. IoT smart building projects have longer timelines from design to build to occupancy, and technology advances are outpacing these timelines. Return on investment models are speculative at best because there are so few implementations to date, and there is little time to conduct pilot or prototype testing.

How smart buildings and IoT came together

A major component of IoT is the internet, but as I mentioned earlier, it is not always required. It began in 1962 as part of the **Defense Advanced Research Projects Agency (DARPA)**. In the early 1970s, it evolved into the **Advanced Research Projects Agency Network (ARPANET)**.

In the 1980s, ARPANET gained support from commercial service providers to be used by the public, and from there it evolved into the internet. Basic communications for devices were mostly connected by satellites and landlines. Tim Berners -Lee proposed the framework of the World Wide Web in 1989 and that laid the foundation of the internet. **Global Positioning Satellites (GPS)** were introduced in 1993 with the Department of Defense providing a system of 24 satellites. Privately owned, commercial satellites were placed in orbit soon after and IIoT became much more functional.

Radio-Frequency Identification (RFID) was used primarily as an inventory tracking solution and was a prerequisite for the IoT in the early 2000s. Devices were tagged and computers were used to manage, track, and inventory them. Walmart and the US Department of Defense were the first to have large-scale deployments of this inventory system. This *tagging of things* has evolved from RFID chips to digital watermarking, barcodes, and QR codes today.

Every item that was tagged was also given a unique **Internet Protocol (IP)** address. **Internet Protocol Version 6 (IPV6)** was implemented by major internet service providers and web companies in June of 2021. They agreed to increase the address space on the global internet by enabling this new protocol for their services and products. Based on 128-bit addressing, IPV6 can support 340 trillion addresses – plenty to last many years.

Considered to be the first IoT device – the first *thing* that began the IoT, John Romkey's 1990 toaster, could be turned on and off over the internet. The toaster was wired directly to a computer since Wi-Fi did not exist yet. Quentin Stafford-Fraser and Paul Jardetzky built the Trojan Room Coffee Pot in 1993 in the University of Cambridge's computer laboratory. A picture of the interior of the pot was uploaded to the building's server, allowing an updated image to be viewed online three times per minute.

Transistors were introduced in the 1940s and the computer was invented in 1951. Couple these with ARPANET, mentioned earlier and introduced in the 1960s, and you now have the three components of the digital revolution. In the late 1960s and early 1970s, people started becoming more environmentally conscious. The 1970s energy crisis saw energy prices rise significantly and the green building movement was kicked off with US government legislation for eco-infrastructure.

In *Chapter 2, Smart Building Operations and Controls*, we will explore in more detail the evolution of various building systems, but for now, we will provide a quick summary. The term *intelligent buildings* was introduced in 1981 by United Technology Building Systems to refer to buildings with its HVAC systems, with a self-claim of minimal energy consumption and better building efficiency. The 1980s also introduced us to mobile phones and personal computers.

The 1980s property boom saw a shift toward intelligent buildings. There was a demand that all these new buildings be as efficient as possible. These intelligent building systems lacked connectivity, but by the early 1990s, the World Wide Web and the internet were introduced. Businesses and buildings were focused on energy-efficient buildings and digital transformation.

As the 21st century began, heavy emphasis was placed on the introduction of computerized programs that better managed HVAC systems to lower energy costs and manage a building's operations. The term *smart building* was coined; however, it would take years before the first *smart building* was built. To this point, intelligent buildings relied on computer programs and RFID tagging systems.

For the introduction of IoT in buildings, we need to look at the sensors that were introduced and connected. Temperature, humidity, motion, gas/air, and electrical current monitoring sensors could be considered the first introduction of IoT sensors in buildings. The final step was to connect all these sensors together in an IoT network and develop software to monitor, manage, and control them.

Figure 1.1 highlights the many applications where IoT sensors and controllers are used to create a smart building.

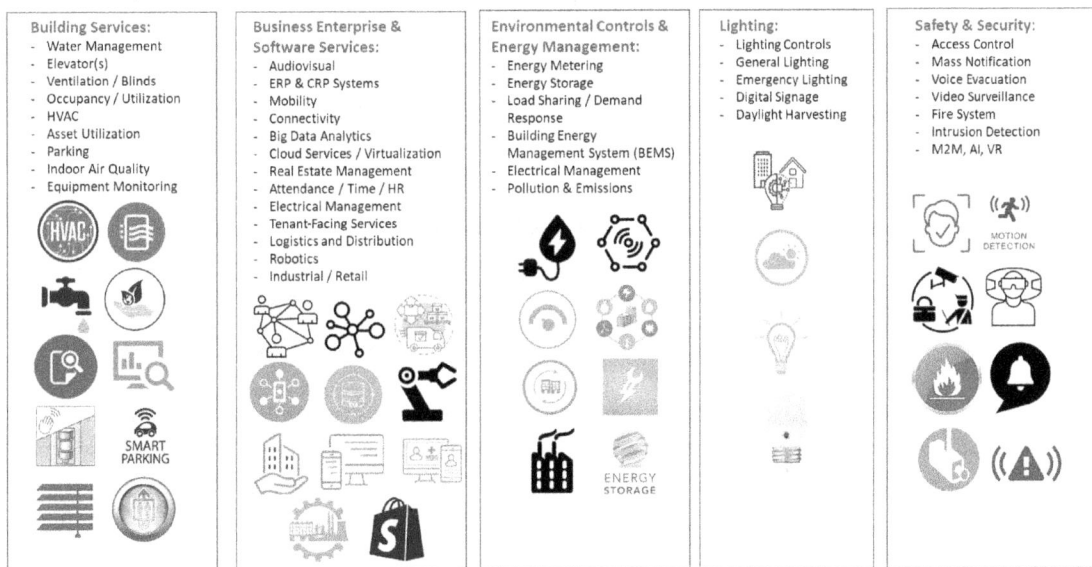

Figure 1.1 – IoT applications in today's commercial buildings

The introduction of IoT provides building owners and operators the opportunity to quickly transition their conventional building to a smart building by simply adding IoT sensors and devices, connectivity, and software-driven applications. But technology simply for technology's sake should not be the objective. These must be purpose-driven to deliver beneficial outcomes for each of the many stakeholders that are reviewed in the next section.

The stakeholders and benefits of a smart building

To best understand the benefits of smart buildings, we need to first understand who the stakeholders are since the benefits vary greatly for each:

- **Owners and Financial Communities**: Building owners can range from a single individual, a partnership or small group of owners, an owner-occupied company, and commercial real estate companies, to banks, investment funds, and **Real Estate Investment Funds** (**REITs**). Their common interest is to generate income, increase the value of their asset(s), lower operations costs, gain access to real-time and historic dashboard information, and leverage special energy and tax advantages many countries offer them.

- **Operators of Commercial Real Estate**: In some cases, the building owner will also operate the building on a day-to-day basis; however, in most cases, owners will outsource that responsibility to companies that specialize in this area, such as commercial real estate companies or property/building management companies. Their focus is on safety, operations, efficiency, cost controls, leasing, quicker response times, accuracy, monitoring and managing building systems, and customer satisfaction.

- **Occupants/Tenants/Visitors**: These are residents in an apartment complex, employees of the owner-occupied company, lease tenants and visitors in a commercial office building, medical staff and patients at a hospital, attendees in a place of worship, or teachers and students in a school. Their common interest is safety, security, comfort, information, direct controls, and their overall QoE.

- **Suppliers/OEMs/System Integrators**: This group includes organizations that contribute or support any of the other stakeholders – **Original Equipment Manufacturers** (**OEMs**), system integrators, cleaning crews, maintenance crews, and numerous others. They are interested in improving the efficiency of delivering their products and services while expanding their offerings and revenues.

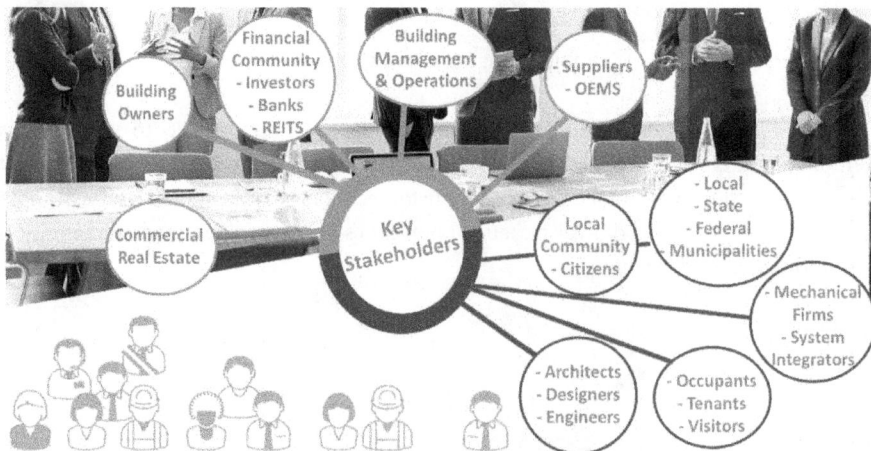

Figure 1.2 – Smart building stakeholders

- **Local Community/Citizens**: This group includes the people that may pass by or look at the building, neighboring buildings, and citizens of the community that benefit from the service(s) the buildings provide (that is, jobs, tax revenues, space, and access). Beyond the use purpose of the building (that is, school, office, hospital, hotel, etc.), these smart buildings can give back extra energy to the grid, reduce the carbon footprint, provide information, and contribute to the overall smart city objectives.

- **Government**: All levels of government, whether it be local, state, or national, are smart building stakeholders. Beyond the similarities to the citizens' benefits, the various government levels want to achieve energy-grid-give-back, green and sustainable buildings, carbon footprint reductions, and the added tax revenues smart buildings bring. Since the government owns and operates many buildings itself, it will achieve operational, energy efficiency, and performance improvements similar to the owners outlined earlier.

Smart buildings operate on a scale; therefore, some buildings will have just a few sensors while others will have many. Building size will also vary, as will the type of building and its purpose. Regardless of size, smart buildings provide digital data about the building itself and the activities within it. The following areas are overarching benefits regardless of building type.

Safety and security

Making sure your building occupants are safe and secure is one of the most important aspects of providing a smart building. Smart building technology can help manage numerous safety-related aspects of facility operations, for instance, remote monitoring of emergency lighting, remote access control, and smart fire prevention systems. In a post-pandemic world, safety starts with IoT sensors everywhere to keep people moving and in a touch-free environment with door access and visitor registration.

Occupancy sensors and wayfinding sensors can assist with safe distancing. IoT within smart buildings supports the functionality of security cameras and fire and alarm systems. IoT air sensors measure, monitor, and control indoor air quality requirements. HVAC automation and controls ensure an automatic supply of fresh air when carbon dioxide levels exceed acceptable limits. Indoor positioning and asset tags help to manage and secure assets. Public safety officials have access to digital information for the building and a reliable communication system that will assist in digitally locating individuals in the case of an emergency.

Cost reduction

Building owners and operators are continuously looking for ways to reduce costs and fend off rising costs to improve the bottom line. The use of IoT sensors in smart buildings greatly contributes to simplifying or eliminating tasks and automating others to lower costs.

Take the case of a tenant in a non-smart building wishing to have the temperature changed in their office suite. They would call the maintenance department, who would open a ticket, and a building engineer would be deployed to the suite to manually change the temperature. This could take 45 minutes or more. With IoT sensors in smart thermostats coupled with a tenant smartphone app, this could be handled directly by the tenant in under 30 seconds.

Energy costs are the most expensive line item after personnel/staff expenses. Smart buildings make it possible to achieve greater savings while simultaneously enhancing occupants' comfort. Energy management and HVAC systems typically are the largest benefactors of smart building IoT, seeing cost reductions anywhere from 10% to 35%.

Smart buildings use information and communication technologies to connect building operating systems together, allowing managers to optimize operations. They also provide greater visibility and control around energy usage and consumption. Facility controls and building operations can all be monitored, controlled, and optimized remotely with the use of IoT. Building management systems, HVAC systems, and all other building systems can benefit from smart building automation and optimization using IoT devices. Predictive and preventative maintenance, inspections, and compliance are cost reduced with the use of IoT. Smart lighting systems and occupancy sensors can lower the cost of electricity. IoT also makes potential problems and issues easier to identify, diagnose, and manage.

Revenue generation and increased asset value

Maintaining and improving lease occupancy rates is one of the highest priorities for building owners and operators. Using smart building IoT solutions helps to differentiate the building from others to help attract and retain tenants and to increase the value of the building asset. New tenant amenity service offerings provide opportunities to charge tenants more. Sustainability initiatives increase the building's value while reducing its carbon footprint.

MIT Center for Real Estate and Real Estate Investment Lab published a pre-pandemic journal article reporting that smart buildings were getting a 37% premium in rent and that their transaction rate went up by 44%. Most companies focused on delivering smart HVAC and energy systems claim reductions from 15% to 40% in energy use and costs.

Improved quality of experience

Often tenants list a comfortable work environment as one of the top criteria for their office. People want to be able to control the airflow around their seating area and the color, tint, and direction of their lighting.

Smart buildings and IoT can greatly improve the occupant's satisfaction with the building. Providing more information, such as real-time IAQ results, wayfinding information, and digitally broadcast information, is high on occupants' want lists. A long list of amenities improved by IoT includes direct access to parking, workout facilities, cafeterias, marketplaces, and other location information and scheduling.

Enterprises and companies located in smart buildings can benefit from asset location and tracking, navigation, wayfinding, real-time occupancy, workforce applications, and the analytical data that can be sourced from each of these. Improved efficiencies, healthy workspaces, and cleaner air can increase productivity and reduce sick days.

Traditional building issues solved by IoT and smart buildings

IoT and smart building solutions can help solve many of the issues facing building owners and operators today. While the building industry has been traditionally slow to adopt technology solutions, recent worldwide health events and the explosion of IoT sensors and devices are helping speed up adoption.

The *new normal* in a post-pandemic world will require permanent changes to create healthy buildings. Hands-free access and devices, IAQ monitoring, occupancy sensing, space cleaning, UV lighting, and space management IoT smart building solutions will remain in place for many years, if not forever. Smart, healthy buildings will be a requirement with IoT sensors and smart applications helping owners and operators to visualize mitigation efforts in real time.

Multiple vendor solutions using proprietary, disconnected systems that create *vendor lock-in* are now being connected with IoT and IP solutions, allowing buildings to use best-in-class products and solutions together at a much lower cost. New non-proprietary IoT network solutions easily connect directly to a building's communication protocol (that is, BACnet) or become the communications protocol in lower-class buildings that do not have a communications protocol today. These IoT and smart building solutions are creating differentiation between buildings and increasing the value of assets in a competitive environment.

Workflow management and work order IoT solutions solve the need to constantly dispatch expensive engineers to perform occupant-requested tasks that they can now perform with their smartphones. IoT sensors and monitors allow building operators to monitor, manage, and control multiple buildings at once from a remote **Virtual Network Operations Center** (**VNOC**). Smart predictive IoT solutions allow for multiple maintenance tasks to be scheduled together to prevent system failures and outages.

Government and tenant pressure to optimize buildings for financial and environmental reasons are being eased with IoT solutions in energy management, operational efficiencies, and occupant-facing solutions. IoT solutions can be used to validate that government-funded facilities are up to standards and are properly maintained.

Buildings are being virtualized with consolidated seamless control systems using SaaS-based technologies to solve these problems as building systems move to the cloud/virtualization.

Smart buildings for a smart city

For nearly 2 years, I spent time traveling from city to city and from conference to conference looking for a magical blueprint for building smart cities for my clients. I discovered that communities trying to develop smart cities faced many obstacles, confusion, and limited attempts with few wins. Trying to define *smart city* was a challenge and each city had widely differing definitions.

Many cities thought free public Wi-Fi and smart lighting were the starting points, but once the vendor-sponsored first few city blocks of smart lighting were completed and city spending priorities were elsewhere, these projects stalled. What I began to recognize was that the path to a smart city began with the buildings themselves and I penned the article *The Smart Way to Smart Cities Begins with Buildings*.

City leaders and city planners have struggled for years to balance their desire for smart cities with the many other needs of their communities. Clearly, building a smart city will help them resolve some of their challenges, such as improving government services, quality of life, energy efficiency, cost reduction, and sustainability, to name a few. These must be balanced with other pressing issues, such as homelessness, urban growth, resource requirements, and decaying infrastructures. Challenges have resulted in limited smart city projects to date and vendor-sponsored initiatives that, while achieving desired outcomes such as smart lighting, typically only cover small sections of the city, and funds would be required to extend these initiatives to the rest of the city.

Smart buildings offer an opportunity for every city to achieve its goals. Self-managed smart buildings can quickly grow into smart campuses and then smart communities by connecting and sharing services. This scalability can be integrated with the city's smart city objectives.

Whether buildings serve as schools, hospitals, or offices or medical, hospitality, residential, or industrial purposes, they are in essence small cities unto themselves delivering the same infrastructural functions as cities do. Safety, security, energy, utilities, lighting, communications, ventilation, sanitation, and parking are just a few of the similarities. When these functions are made smart in a building, they can help a city transition similar functions to smart to build a smart city. Basically, a building is a microcosm of a city; therefore, a smart building is a microcosm of a smart city. Connecting smart buildings together can build a smart city foundation.

Buildings and cities must work together to achieve mutual goals. Energy is one of the biggest opportunities and as smart buildings become more energy efficient, they may also be able to give back to the city's energy grid. IoT sensors and connectivity with the grid make buildings more responsive to grid conditions to reduce stress and improve reliability by cutting energy consumption during high-demand periods. Grid-responsive equipment such as water heaters turn on and off in response to the utility's peak demand. Buildings can also collect and store energy with solar panels and batteries.

By collecting and analyzing data, safety and security is another area where smart buildings can help build smart cities. With building-mounted cameras and IoT sensors such as gunshot and occupancy sensors, information can be shared with the city's integrated control center. Suspicious activity observation, traffic management, and crowd control can all be managed centrally. Analytics and machine learning can leverage historical data to predict situations and trigger alarms if needed.

Smart city emergency response starts with smart building information. Police, firefighters, emergency medical responders, and other first responders can access digital information about a building, such as floor plans, what chemicals might be stored there, and what building systems are in use. In the case of a building fire, remote access and control of the building's management, HVAC, and fire suppression systems can ensure that the proper amount of air is supplied or cut off as needed to manage the fire while the first responders are en route. Video cameras and motion sensors can provide valuable insights and information prior to arrival so they know what to prepare for and what to prioritize. Real-time information regarding the situation can be transmitted directly to hospitals, police, and firefighter support teams.

A building is a physical asset that, when enabled with IoT sensors and software, transforms into a *smart asset* managing its internal systems more efficiently. These smart assets connected to other smart assets and smart systems become part of the larger city ecosystem. Smart buildings are an important part as they provide a variety of network connectivity options while serving as a platform for other sensors and devices that can capture, share, and communicate with each other. For example, smart buildings can be linked to police departments via IP-connected video and to on-street parking via sensors, as well as to smart outdoor lighting systems, and all of these can function together to provide a responsive and safe environment for residents, businesses, and visitors.

Many of the buildings within a city are government-owned and managed buildings such as courthouses, libraries, fire stations, police stations, schools, community colleges, and others that make up the city's infrastructure. To build a smart city with smart buildings, it makes sense to incorporate smart requirements into these publicly funded buildings, such as mobility, healthcare, security, lighting, environment, energy, construction, and communications requirements.

Data is collected from almost every smart device and sensor located in, on, or around a building. Public data and information from smart buildings combined with data collected by the government and other sources can be analyzed to solve city problems and make improvements. While, to many, data gathering feels like a privacy invasion, it has become an integral part of life. Experts around the world are constantly creating new solutions and programs to reduce the risk of data breaches. Government entities along with industries are creating privacy guidelines such as the **General Data Protection Regulation** (**GDPR**) in Europe.

The history and evolution of building control systems

The key to leveraging all these IoT smart building solutions lies in the ability of building owners and operators to unify their legacy building systems with new controls, sensors, and IoT devices for real-time, seamless access, management, and optimization. The evolution of smart buildings didn't happen overnight; this has been slowly developing over many decades.

Prior to the 1970s, building management systems were local, with simple pneumatic controls. Pneumatics uses pressurized or compressed air that is distributed down a main line to control devices connected to that line. Air leaves through what is called a branch line and these branch lines act as a control signal to a device such as a thermostat and its controlled air damper actuator.

While pneumatics was still in heavy use in the 1980s, analog electric controls were introduced. These simple controls worked by turning a knob that injected resistance into a circuit. This resistance triggered the control device (valve, relay, etc.) to react. Electromechanical control systems were combined with pneumatic systems to control devices. The element would expand or contract on a thermostat that would open or close a circuit to turn the unit on or off. These systems were prone to calibration issues.

This was followed by the introduction and use of microprocessors, computers, and distributed digital process controllers in buildings in the 1990s, dubbing it the era of centralized controls. Direct digital control systems are still used today. Software programs were written allowing technicians and operators to control sequences by changing code. Controllers were daisy-chained together, creating a wired network. **Building Automation Systems (BASs)** and **Building Management Systems (BMSs)** were introduced. (We will review these systems in *Chapter 2, Smart Building Operations and Controls*.)

Fast-forward and the 2000s saw the intelligent buildings era, with the introduction of common in-building communication protocols such as BACnet and LonWorks. These protocols allowed individual devices to communicate with a central building system. Distributed digital computers were located on individual devices and communicated with the central system. Databases were created that facilitated analytics and the growth of energy management systems.

Over the last decade, intelligent buildings have transitioned to the *smart buildings* of today with central controllers communicating with powerful cloud-based software and AI-based machine learning applications that can optimize building designs, conserve energy, and predict equipment failures before they happen, demonstrating a vast improvement to building management overall. Unfortunately, in 2023, these systems are still disconnected, and sometimes proprietary, leading to multiple applications, silos of data, and user frustration.

I imagine that the next evolution of the smart building will be the *unified building*. A unified building fully connects and integrates all systems, components, sensors, and devices on a single platform to allow access and control and it will provide a unified view of the building on a single pane of glass. This will provide full integration of energy, facility, IT, security management, and control systems on a comprehensive, unified platform. Fully integrated and connected microprocessor-based controls and sensors will deliver massive amounts of data utilized for ML and AI applications.

Summary

This introductory chapter provided you with a view of what it might look like to work inside a smart building and the many productivity, comfort, and efficiency improvements that IoT smart buildings can deliver for the occupants. We established an understanding of what the IoT and smart buildings are and then we connected them together to see the endless possibilities for smart building applications.

These IoT smart building applications solve many of the challenges faced by building owners and operators today. They deliver safe, efficient buildings that are energy efficient, reduce operations and energy costs, increase the value of the asset, and deliver a greatly improved quality of experience for the occupants. All buildings, regardless of function (schools, hospitals, retail premises, hotels, offices, factories, etc.), can achieve these benefits. Since buildings are a microcosm of a city with many similar challenges and requirements, the way to smart cities will need to start with smart buildings.

We touched on a brief history of how building management systems evolved from pneumatic controls to the smart buildings of today. Since a building is made up of a number of systems, the next chapter will review these key systems, their functionality, and how each of these will greatly benefit from IoT smart building applications that you can design to create a smart building.

2

Smart Building Operations and Controls

A building is comprised of many systems, each operating independently, and most are not connected or communicating with each other. These siloed systems include mechanical, lighting, electrical, energy, plumbing, ventilation, heating, and air conditioning, to name a few. In *Chapter 1, An Introduction to IoT and Smart Buildings*, we imagined what it would be like working in a smart building and discovered smart solutions in almost every function of the building's infrastructure comprised of these systems.

Building a smart building starts with making some or all these systems smart with IoT sensors, actuators, controllers, and devices. They are then connected via a communication network, and data is delivered to a computing platform located on-premises or in the cloud. These systems are monitored, measured, controlled, and optimized using smart building solutions. Data analytics, machine learning, and artificial intelligence computing can then be performed, and commands can be delivered back to each system or across systems. Once these systems are integrated, building automation and predictive solutions can be developed and implemented.

This chapter will introduce some of the key building systems and describe their functional responsibility and relationship to other systems. From there, we will highlight the opportunities and benefits that can be achieved by introducing IoT technology to each system as you build your smart building.

In this chapter, we're going to cover the following main topics:

- Building operations and maintenance and what benefits can be gained from introducing IoT technology.

- Remote monitoring and management, especially during COVID-type outbreaks, and reducing operations costs.

- **Building Management System (BMS)** and/or **Building Automation Systems (BAS)** and understanding the major role they perform in managing and controlling other systems, and why they are critical to building a smart building.

- Building energy management systems that perform monitoring and metering to collect energy data and develop insights on a building's energy usage.

- **Heating, Ventilation, and Air-Conditioning (HVAC)** systems, which consume 40 to 60 percent of a building's energy and are well suited for IoT smart building solutions.

- Most buildings begin their transformation journey to becoming a smart building with a simple lighting project. We will explore how these IoT projects deliver immediate results and improve occupant satisfaction.

- The ability to introduce IoT into the facility request and management process is another relatively easy way to begin the transformation into a smart building.

- Maintenance costs and the downtime associated with system outages can be greatly reduced by IoT solutions and data management to predict replacement times.

- Introducing IoT to accurately bill tenants for use of HVAC and lighting during non-business hours.

- IoT sensors contributing to setting up monitoring solutions that can then push notifications to building engineers to improve response times.

- Space management and how IoT smart building technologies can greatly ease constant space planning requirements.

Facility controls/building operations maintenance

Managing a building requires specific skills and tools for each of the functional areas, such as building operations, maintenance, engineering, forecasting, budgeting, health, safety, and security. To ensure a building is functioning as it was designed and built, several competencies, services, processes, and tools are required for the day-to-day operations. An effective maintenance process ensures that critical assets are in good condition, resources are allocated efficiently, procedures/schedules are enforced, and building performance is managed.

Operations and Maintenance (O&M), or operational maintenance as it is commonly referred to, are the day-to-day activities necessary for a building and its systems, equipment, and occupants to perform their functions. O&M includes the maintenance of the physical building itself, management of the building's systems, landscaping, groundskeeping, site improvements, and maintenance of the furniture, equipment, and the building's interior.

Reactive maintenance is the approach of running equipment until it fails or breaks and was the traditional methodology used. According to a 2021 PE Maintenance report, reactive maintenance is still practiced in over 51% of the buildings today. IoT sensors and other smart building technology solutions have been developed for O&M activities to optimize operations efficiency, extend equipment life, reduce capital repairs, reduce unscheduled shutdowns, and lower costs.

Work orders are at the center of any O&M system. They are a simple tool to define the task that needs to be completed, for scheduling, assigning, and tracking job tasks to completion. Digital work order systems allow work orders to be completed on a smartphone. The status of the workflow is updated at each step using IoT, barcode, QR, and other smart sensors. Work orders are used to provide real-time statuses and prioritization, and they provide a digital record that can be used for invoicing, audits, and historical recordkeeping.

Asset management is considered one of the top IoT use cases in buildings today. Physical devices are tagged, scanned, and connected to asset management software and linked to databases. Important information about the asset can be viewed and updated using a smartphone or smart device. Where the asset is located, who is using it, who has responsibility for it, the condition of the asset, its maintenance history and schedule, and even the user manual can all be monitored and stored in one asset management system. Tracking tags located in or on the asset are considered an IoT device or sensor. They collect, store, and communicate the asset's information.

Computerized Maintenance Management Systems (**CMMSs**) consolidate the procedures and practices used to track a building's systems maintenance information. Detailed descriptions, historical data, schedules, logs, and costs are typically managed through the CMMS. These maintenance activities are typically repetitive and include scheduled, preventative, and emergency maintenance activities. O&M manuals outline the processes, methods, components, tools, and schedules for the proper maintenance of a physical asset.

Computer-Aided Facilities Management (**CAFM**) began as basic space planning but now has evolved to include CMMS, **Building Information Modeling** (**BIM**), and an **Integrated Workplace Management System** (**IWMS**). An IWMS is a software package that helps manage the infrastructure and facilities assets along with the associated software. BIM is the process and use of technology to create a digital representation of a building, its spaces, and its functionalities.

CAFM helps to plan everything for a building's day-to-day operations. Applications may include space planning, asset management, move management, maintenance management, room reservations, facility operations, and customer service requests. With IWMS, operators can manage purchases, leases, finances, and sales information. Capital projects for remodeling or development of new facilities are managed here.

It combines architectural, engineering, business administration, and construction concepts to optimize running the building. What differentiates CAFM from other information technology solutions is the use of **Computer-Aided Design** (**CAD**). It provides a visual representation such as interactive floor plans with other key data/information points.

Janitorial and cleaning processes and schedules ensure the building and all its assets are clean for a healthier building. The use of specific chemicals and the type of surface being cleaned are scheduled and tracked. Certain chemicals may trigger total volatile organ compound levels and impact indoor air quality, as measured by IoT environmental sensors.

IoT devices are used to notify crews of cleaning services required or recently completed – for example, a work area and assets have been cleaned and the scheduling software has been updated to allow the next person to use the space. Housekeeping, groundskeeping, landscaping, janitorial, and custodial services are included as general maintenance activities.

Proper building operations and maintenance improve the safety of a building, reduce the cost of capital repair, reduce building ownership costs, improve uptime, and improve occupant satisfaction along with the bottom line.

Remote monitoring and management

COVID-19 disrupted facilities management operations and forced facility managers to figure out how IoT technology could help them manage their buildings from a distance. Significant efficiencies and full asset visibility can be achieved along with real-time alerts to provide an overview of information for decision-making. Almost any physical device can be fitted with an IoT sensor and monitored remotely.

Smart sensors help facilities managers to keep properties in good working order and free up engineers to then take on more complex tasks. Other sensors can flag risks, such as burglaries, gas leaks, or poor indoor air quality. Remote dashboard monitoring delivers real-time data to identify issues at a much earlier stage. Some systems use machine learning to automatically adjust and redefine maintenance plans to prolong the life of the appliances.

The benefits of a remote monitoring system are as follows:

- The visibility of a building's performance in real time
- Management and control of building systems remotely
- A reduction in cost and identification of energy waste
- A reduction in the carbon footprint
- The need for manual work removed
- The chance of human error removed
- Automated compliance and reporting
- You remain aligned with legislation

Almost any system within a building is a candidate for IoT sensors and devices used to monitor, manage, and control a system. These include HVAC system monitors, security monitors, fire/smoke detection, water leak detection, temperature, humidity and environmental monitors, and cameras. Most sensors and devices have the capability to send real-time email and/or text alerts to a property manager. Multiple buildings can be monitored simultaneously in a **Network Operations Center** (**NOC**).

In a post-pandemic environment, IoT sensors and video analytics can identify non-compliance or risky behavior in terms of social distancing compliance and occupancy measurement to comply with COVID-19 protocols for reduced occupancy. Thermal sensors can measure body temperature to determine whether entrance should be denied. Zero-contact solutions such as touch-free elevator buttons and access tags help to reduce the risk of viruses spreading and can help keep intruders from accessing a building. Contact tracing through tracking occupant movement can limit the spread of contagions. IoT sensor technology can be used to optimize office space by notifying managers when a desk becomes available.

Commercial and residential tenants may own or manage environmentally sensitive products such as temperature-sensitive food and beverage products, and humidity-sensitive products such as paintings and other artwork. Remote monitoring services can ensure the conditions are within the acceptable thresholds for these products to ensure their quality and safety.

Configuring alarms requires access to in-building systems that are complex and difficult to use. In many cases, it is not possible to see across systems, buildings, or portfolios in a consolidated dashboard. Remote building monitoring systems continuously monitor device alarms, send an alert when points are out of range, send a notification when the ranges return to normal, and collect historical points data for analytics. Typical features include the following:

- Identifying building devices and set points
- Creating normal in/out range thresholds
- Setting alarm priority and risk level
- Designating each monitor
- Adding subscribers to the notifications
- Sending real-time alarm emails and/or text messages to subscribers
- Providing historical device data using the dashboard
- Resetting or adjusting monitoring thresholds

Investing in remote monitoring and management does require an upfront investment. As the cost of sensors stabilizes post-normalcy in the semiconductor supply chain, as does inflation, the long-term investment can be offset by new efficiencies and cost avoidances by keeping equipment running within its proper ranges. Once the monitoring network's basic networking and communicating capabilities are in place, you can start with a few strategic sensors to monitor critical areas such as security, energy, fire safety, and building access. Additional sensors can be added quickly to add additional monitoring capabilities. These systems can collect and analyze massive amounts of data to better understand and manage a building.

Figure 2.1 depicts just some of the systems within the building that are candidates for the implementation of IoT sensors and devices.

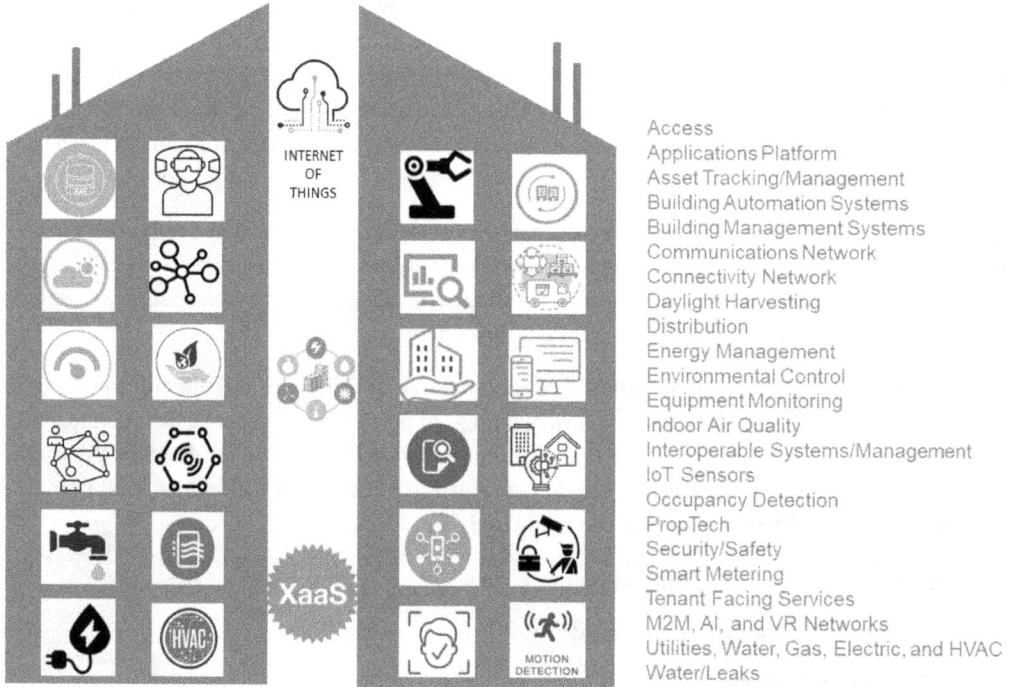

Access
Applications Platform
Asset Tracking/Management
Building Automation Systems
Building Management Systems
Communications Network
Connectivity Network
Daylight Harvesting
Distribution
Energy Management
Environmental Control
Equipment Monitoring
Indoor Air Quality
Interoperable Systems/Management
IoT Sensors
Occupancy Detection
PropTech
Security/Safety
Smart Metering
Tenant Facing Services
M2M, AI, and VR Networks
Utilities, Water, Gas, Electric, and HVAC
Water/Leaks

Figure 2.1 – A building system of subsystems

Later on, in *Chapter 10, Understanding Your Building's Existing Smart Level and Systems*, we will present a number of different assessments that can be used to determine the current state and readiness of your building. An exhaustive list of possible systems will be provided to help determine your building system's subsystems.

Building management systems/building automation systems

Almost every sizable building regardless of its use will have some form of technology that is used to connect and manage the building's control systems. These computer-based control systems monitor and manage mechanical and electrical equipment, such as energy management, HVAC, and lighting systems. The control system is commonly referred to as the **Building Management System (BMS)**, or the **Building Automation System (BAS)**. While there is a technical difference between BMS and BAS, the terms are used interchangeably, and the differences are very minor. A BMS is focused on monitoring and maintaining building operations with supervisory control, while a BAS is typically a subset that is designed for the automation of building systems with limited energy-efficiency capabilities.

The goals of these computer-based systems are to improve system uptimes, improve system efficiency, reduce costs, and improve safety and occupant comfort. IoT, automation, and advanced analytics help building operators reach their building performance goals. A traditional control system will be comprised of servers, sensors, controllers, field buses, inputs, and outputs.

These control systems provide real-time monitoring, trending, scheduling, controlling, and logging of a building's operation and the performance of day-to-day operations. They monitor and control zone temperature, air volume, air quality, air handling, and exhaust fans during and after normal building hours. They measure and monitor the building's performance and provide equipment alarm and fault notifications. These control systems interact and connect with other building systems.

Generally, a control system will have three major levels. The first level is the field level and consists of the actuators, valves, sensors, and thermostat devices that deliver the basic information or inputs. These inputs are sent to the automation level, which consists of programmable modules and relays that are set to threshold points. The third level is the management level, whereby the information is displayed in a graphical format. Monitoring and control are performed at this level.

Control systems are adaptable to change, and new technology and equipment can be easily added and integrated with existing systems. Buildings with control systems typically have a higher asset value than those without. While these systems help simplify the management of a building, reduce operations costs, and can protect equipment functionality, there are potential drawbacks as well.

These systems are very expensive, sometimes costing hundreds of thousands of dollars. Data limitations such as proprietary languages and lack of standardized naming conventions can prevent us from achieving maximum savings and efficiencies, and other systems are then required. Many times, smaller equipment is not connected, and potential savings opportunities are not achieved from these.

Finally, these are siloed and disparate systems that often have their own proprietary languages and do not work well with other systems. IoT solutions and analytics not only collect data from these systems but also automate and control actions for just about every aspect of a building.

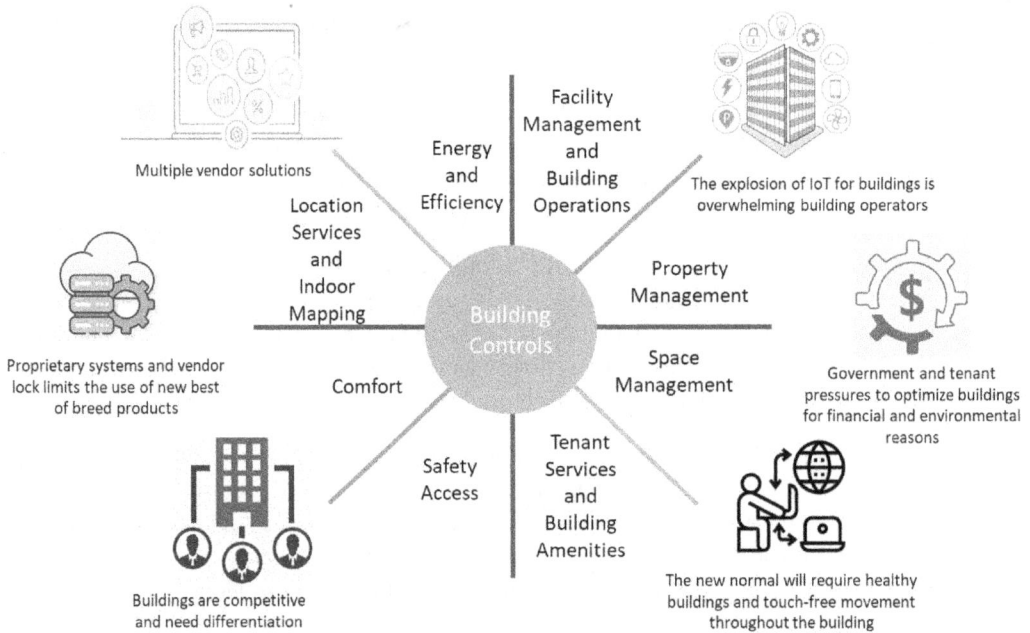

Figure 2.2 – A building's control system challenges

A recent study by Technavio forecasts that the building automation and control systems market will increase by USD 31,125.12 million between 2022 and 2027, which is a **Compound Annual Growth Rate (CAGR)** of 9.25%. This growth continues to evolve, with new products largely driven by energy efficiency requirements, security systems, and comfort requirements. Major industry leaders include Siemens, Honeywell, Schneider Electric, Distech Controls, Delta Controls, Johnson Controls, and Carrier; however, new smaller companies are beginning to disrupt the industry with innovative IoT solutions and platforms.

Building systems are enablers to deliver outcomes. These outcomes deliver solutions that solve challenges for a building owner and operator while achieving business and financial goals. Historically, these systems shared separate domains and did not share data or controls. IoT and smart building technologies are driving the convergence of these systems while developing new operational technology stacks.

Connecting these systems, devices, sensors, equipment, and applications produces data across different boundaries, stakeholders, and capabilities to build a centralized management platform control for a building. Access, HVAC, lighting, metering, sensors, equipment, video cameras, and other systems are interoperable under a single management architecture.

The remainder of this chapter will focus on some of the key BMS/BAS subsystems or connected systems.

Energy management

Improving energy management is great for the environment and great for a building's bottom line. Energy costs typically account for 40% of the building's total operating budget and, therefore, easily become the most common targeted area for improvement and cost reduction. Worldwide sustainable goals focus on carbon footprint reduction, climate control, and renewable energy sources, and buildings contribute enormously to each of these. Efficient energy use in buildings can decrease indoor air pollution and reduce health-related problems.

A **Building Energy Management System** (**BEMS**) is a computer system that monitors, controls, measures, and optimizes energy consumption within a building. A BEMS will specifically connect a building's systems, which involve energy use and demand such as electrical and mechanical lighting, HVAC, ventilation, power, and security systems. Metering and sub-metering capabilities allow facility operators to collect important energy usage information. A BEMS is typically part of a BMS/BAS and the majority of its work is done at the management layer.

Building energy consumption management occurs in three ways. The first is energy conservation by eliminating ways in which a building wastes energy. The second is energy recovery by reusing the byproduct of one building system's source of energy for another system – for example, wasted heat from a manufacturing process could be used to heat the building's water. The third method is energy substitution, whereby the normal source of energy is replaced with a more economical and often more environmentally friendly source.

A BEMS is often part of a BMS/BAS but may at times be a separate system. Either way, the energy management process starts with sensor placement throughout a building. These sensors measure temperature, changes in voltage, vibration, spikes in energy usage, humidity, motion, light, smoke, acceleration, chemicals, pressure, and any other change that can be monitored. Sensors can be placed throughout a building, on machinery, in lighting systems, in HVAC and ventilation systems, on hot water heaters and pumps, in refrigeration units, and so on.

These sensors are either connected to a device or part of the device. The device will have the capability to connect to a network, and data can be transmitted to a local computing device or the cloud. Wireless access equipment is now the preferred method and includes IoT terminals, IoT modules, cellular dongles, routers, or gateways. Information from the sensors is collected and managed by software that can identify faulty equipment, sources of inefficiency, and opportunities to match energy production with actual demand.

Actuators act in an opposite way to sensors; whereas a sensor detects, an actuator will act. When they are triggered, they will take action. Valves, motors, and electric switches are actuators. Smart metering tracks resources that are being distributed and consumed. Precise energy accounting and demand forecasting can help streamline operations. I/O voltage and current readings, water flow, pressure, and consumption trends are provided.

An application will be on a device itself and provides logic, such as when a temperature exceeds a preset threshold, a message is sent to a network. The application will run on a processor that is called either a **Microcontroller Unit (MCU)** or a **Microprocessor Unit (MPU)**. They are connected through a network, and energy management software can be on a local computer or in the cloud.

Since a building's energy consumption continuously changes, efficient energy management requires the right data at the right time. This data can show occupants' behavior and energy usage patterns, along with the total consumption of all the systems and equipment connected to a building. Time-of-use changes, seasonal factors, and real-time weather data can greatly aid in generating proposed solutions. A BEMS with IoT sensors and actuators can provide a holistic view of a building's energy consumption and provide insights to improve energy efficiency.

Lighting systems

For many buildings, the transition to a smart building usually begins with the lighting system. Immediate and significant results can be achieved with reasonable investments in lighting systems, and therefore, they tend to be an easier decision for building owners and operators. Most of the other building systems are scheduled-based systems, often still running when no one is using the building. Smart lighting systems can be occupant-driven and better utilized when occupied or unoccupied.

Smart building lighting systems typically begin by converting to using LED systems and lamps, which consume less energy and reduce costs. Lamps are changed much less frequently, contributing to lower maintenance costs. Most smart lighting systems are IoT sensor-controlled motion sensing systems, which improve safety and convenience. They can also improve productivity, as they are bright enough to light all areas and are less of an irritant to the eye. The data collected helps building operators know which spaces are used most and least, setting the stage for other temperature and ambient changes and integration with other systems.

Smart lighting in conference rooms can be integrated into the conference room scheduling system, and the lights can be flashed to notify the occupants when there are 10 minutes left on the schedule. Numerous other flash notifications can be established, such as package arrival notifications, security or safety breaches, and severe weather warnings.

Power over Ethernet (PoE) allows delivery of DC power between 44 and 57 volts to devices without separate power suppliers via the Ethernet (a twisted-pair copper cable). This can reduce the upfront construction costs and the amount of installation time and effort. It requires less energy and is much cooler to run. Lighting also generates heat, which must be compensated for with the HVAC system. Light sensors can automatically adjust window shades to reduce solar heat gain. Since POE is already paired with data cables, it can be connected to a software control system that manages use and schedules and can be paired with shades and HVAC systems.

Smart light bulbs can be controlled remotely, and the brightness can be adjusted. These smart bulbs can do much more than light up a room. Some integrated luminaires have other sensors built in for temperature, humidity, occupancy, vacancy, asset tracking, way-finder beacons, and even speakers.

Network lighting systems can provide energy savings through multiple control strategies such as daylight harvesting, task tuning (dimming lights based on the use of the space), time scheduling, demand requirements (each space illuminated based on needs), and load shedding (utility company requests auxiliary lighting be shut off). Daylight sensors can dim or even turn off lighting when sufficient daylight is available.

Smart lighting extends beyond the inside of a building. LED lighting on the building's facade can easily differentiate the building from others and make the building stand out in the landscape of a city, while keeping consumption down. IoT sensors mounted with the exterior lighting can detect gunshots, measure snowfall, detect noxious fumes, and house public Wi-Fi hotspots. Parking lots and street lighting help build a smart city infrastructure and improve city security and safety.

Facility support

Building occupants are driving new requirements that are forcing building operators to change their way of thinking about facility support. Millennials are accustomed to real-time demand-driven services fueled by apps such as Uber and online food delivery services. Traditionally driven by cost and not value priorities, facility support is evolving to real-time, connected services that are enabled by the latest technology. These services will drive premium rates, differentiate buildings, and improve the quality of experience for occupants who expect personalized, service-led facilities.

Data-driven facilities management technologies can deliver agile, integrated workplaces by enabling IoT and AI solutions to drive efficiency and improved customer experiences. Connecting and unifying building management services enables workplaces to perform better and to be more interactive. Legacy computer-aided facility management systems will not meet the new requirements; therefore, IoT and AI solutions will create value.

What is facility support? It's comprised of individuals, service partners, vendors, and teams that are responsible for running a facility and delivering services to employees, visitors, and other stakeholders. These include the in-house staff who are there to interact with and support the occupants, waste management services, janitorial services, housekeeping, repair, and seasonal services.

It includes tradespeople and craftspeople such as carpenters, electricians, plumbers, and other specialty crafts. Security services such as guards, security firms, and campus police are also included. From a broader perspective, environmental testing and inspection experts and administrative services are required to keep a building functioning and maintained.

Building managers need to coordinate all these requests and tasks effectively. Digital support work order ticketing systems can be used to route and track requests, while asset management and digital twins can quickly provide needed information such as floor plans and user manuals. IoT solutions will connect and maintain digital systems and, when integrated with facility operations software solutions, will create an integrated facilities operation system.

IoT sensors in bathrooms can notify staff when soap, paper towels, or toilet paper are low and require replacement. Other sensors can detect spills or areas that may require cleaning. Potentially dangerous situations such as a chemical spill can be identified, with alerts sent to managers or connected systems that can respond automatically.

Embedded beacons can track and locate a service employee's location. Other IoT beacons can monitor the state of equipment and verify facility security. These IoT applications will allow managers to unite all functions to monitor and track these operations simultaneously. Many IoT devices for facility operations have yet to be developed, but this area is growing very quickly.

Preventive/predictive maintenance

Nearly every functioning piece of equipment within a building will require cleaning and maintenance at some point. Building and equipment maintenance is very time-consuming and expensive, and knowing when to perform it can prevent outages and maintain costs. In the past, some buildings used a reactive-maintenance approach that allowed the equipment to fail and then perform the fixes. Obviously, this is an inefficient expensive approach that could negatively impact occupant satisfaction levels during downtime.

Some maintenance activities are easy to schedule and perform on a calendar basis, such as once a week, once a month, or once a year. Others can be performed based on the actual runtime, such as the number of miles or hours a piece of equipment has operated since its last maintenance.

Condition-based Maintenance (CbM) is performed once equipment starts showing signs of wear and tear or is operating out of normal range, such as vibration, temperature, or pressure. Condition-monitoring sensors conduct spot checks during normal operations with technologies such as infrared thermography, acoustic monitoring, current analysis/discharge, oil analysis, and vibration analysis.

Preventive Maintenance (PM) is performed proactively to reduce the possibility of failure and downtime while extending the useful life of the equipment. Typically, this is done by inspecting equipment regularly for problems and then repairing it before failure occurs. This will generally help the equipment run more efficiently. Calibrations, lubrications, cleaning, inspections, adjustments, or part replacements are part of preventative maintenance. An asset's condition is documented during maintenance so that future maintenance can be scheduled.

Scheduled preventative maintenance does come with a cost. Because the equipment's condition is not always taken into consideration, some scheduled maintenance could be unnecessary or insufficient. This could result in not enough maintenance or too much. Enter IoT solutions to address these challenges. Through real-time remote monitoring, the equipment's condition can be assessed and managed. An IoT sensor can condition-monitor equipment, and performance metrics can be run with alerts sent when conditions are out of normal range.

Using IoT smart building technology and analytics from data collected, predictive maintenance can be proactively conducted. It is a strategy that has been proven to be more efficient, less costly, and minimally disruptive. It increases maintenance effectiveness and is a better use of time for the building technicians. They can perform other value-adding tasks instead of performing unnecessary routine tasks.

IoT sub-metering provides building operators with greater visibility of individual components to identify any machinery or process inefficiencies. Companies can track their consumption and compare them to competitors to stay ahead of them. Sustainability goals can be established and progress tracked against these.

Predictive Maintenance (PdM) works by adding IoT sensors, artificial intelligence, and real-time smart dashboards. Sensors can be placed on doors, windows, pipes, or pretty much any physical device that requires monitoring. They are easy to install, usually very discreet, inexpensive, and very effective in providing real-time data.

Large amounts of data coming from these sensors allow a facility to correlate the data with other data sources to establish a baseline and desired performance figures. Machine learning and AI drive predictive maintenance computations using data and desired performance information. Real-time smart dashboards can be delivered to smart devices with real-time actionable insights in one building or across a portfolio.

PdM is usually cost-effective, but ROI calculations should be performed to justify spending money on non-mission-critical equipment. The amount of data collected by sensors provides precise asset information to repair equipment and eliminate downtime. PdM increases equipment's service life and improves the overall quality of a building and its equipment. PdM systems typically pay for themselves within a few years.

Many building owners and operators have developed a balanced approach that assesses an asset's conditions and determines maintenance requirements in relation to the impact of its operation. Not all equipment is of equal importance, and maintenance can be prioritized and operations optimized over time. *Figure 2.3* indicates some of the systems important to building operations, such as all the **Building IoT** (**BIoT**) devices, and elements important to tenants and occupants, such as air quality and temperature:

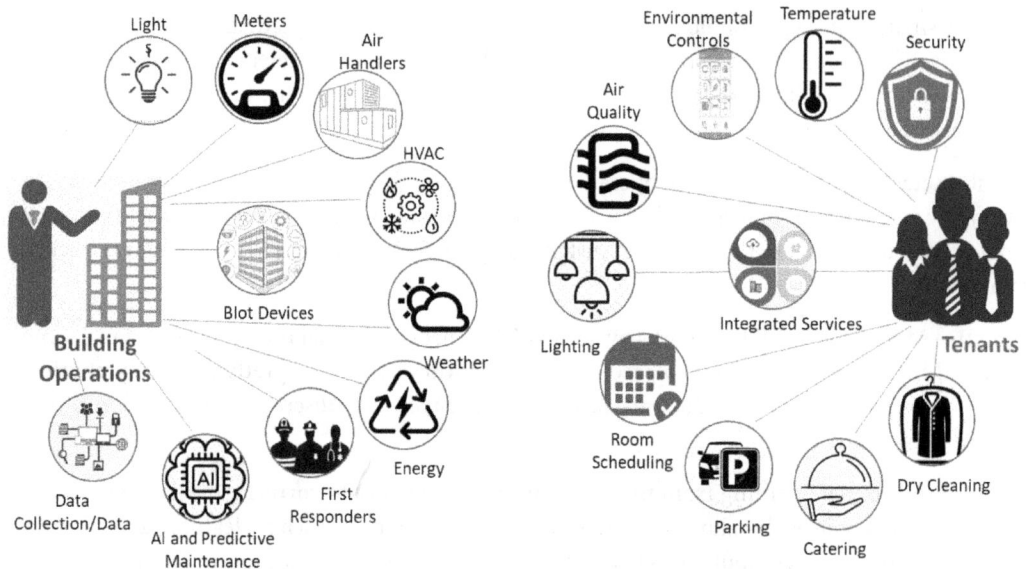

Figure 2.3 – A building's operations and tenant services

Proper maintenance can extend the life of an asset, increase equipment uptime, and help to keep occupants satisfied. Another tenant satisfaction trigger is the proper billing for utilities, and sub-metering is a method to improve accuracy.

Tenant billing for sub-metering usage

Sub-metering tenant billing creates accountability and ownership of utility use by commercial or residential tenants. It ensures fair and equitable billing for actual gas, water, and electricity use, as opposed to a straight, across-the-board split based on prorating per square foot.

Tenants can monitor and better manage their use, making them more likely to use less and giving them satisfaction. In the end, the property owner will achieve cost savings while extending the power supply equipment lifespan. Adding sub-meters adds value to a property and tends to attract a more qualified tenant.

So, what is sub-metering? In the past, utility usage for a building was measured on one master meter, which provided a reading for an entire building. Building managers spent numerous hours calculating the monthly bills for each tenant, often containing billing errors and creating confusion. Sub-metering simply places meters on each floor, for each utility and individual unit. This allows for accurate readings and billings for each tenant for actual utilities they consumed.

Sub-metering has been around for years but is becoming easier to implement and more affordable with IoT sensors, software, and other technologies. Detailed information allows a tenant to have better insights and ideas on how to reduce usage. IoT sub-metering is made possible by IoT sensors connected to existing meters. These sensors measure the rate of water, energy, and gas usage along with the number of pulses that are generated by the meter.

An **Automated Tenant Billing System** (**ATBS**) uses IoT-based sub-meters to automatically collect and accurately bill tenants. These systems can integrate with existing BMS systems or can be standalone IoT hardware to collect data. They automate the billing process with actual usage, based on the tenant's space and allocations for common areas. Other systems can use QR codes. Tenant portals and apps let tenants monitor and control their energy use habits and forecast for budgeting.

Sub-metering with IoT solutions improves the accuracy of billing, provides real-time consumption data, and helps companies make informed decisions to optimize utility use. It can provide information used for PM and PdM and repair to prevent equipment failures. It encourages tenants to shift their usage to off-peak hours during peak demand when utility costs are their highest. Tenants can have direct real-time access to their utility usage on their smartphones.

Maintenance alerts/notifications

Earlier in this chapter, we reviewed PM and PdM IoT smart building applications, whereby maintenance alerts can be sent to building operators for action. Alerts and notifications may also be required for events not tied directly to maintenance activities such as sewage/drainage blockage, water, gas, fluid leaks or spills, and weather-induced incidents such as winds, wildfires, or flooding.

Detecting incidents and anomalies as quickly as possible is critical for safety and for limiting damage, downtime, and costs. IoT sensors and networking communication technology ensure that incidents are detected and communicated quickly.

Compact wireless sensors can measure most physical items nearly anywhere. Intelligent monitoring solutions and software can create rules that direct when an alert should be created and sent. Connected networks can communicate information anywhere via any method such as text messages, emails, flashing strobe lights, digital displays, or phone messages.

Beyond these alerts, there is often a need for notification to key stakeholders. In some incidences, immediate emergency evacuation notification may be required. For example, IoT gunshot sensors in or around a building can trigger immediate notification to occupants and emergency services of an active shooter situation. IoT sensors can detect smoke, smells, chemicals, pressure, and numerous other items that may trigger responses and notifications.

Smart immersive glass surfaces and screens can display emergency action plans and real-time evacuation maps. Broadcast text messages can be sent, or a notification can be delivered directly from a building's smartphone app that is used by all tenants.

Building-wide or even campus-wide notifications may need to be sent for widespread destruction, outages, or scheduled maintenance notifications.

Space management

The return to buildings post-pandemic created a major focus on space occupancy and space management. Social distancing guidelines, hybrid-working arrangements, and continuous virus-variant outbreaks have building space planners searching for technology solutions. By combining IoT sensor data with advanced analytic solutions and space management software, we can deliver actionable solutions in many areas to save money and enhance operations inside and around buildings. Workspace changes could be made in real time by analyzing vacancy rates, utilization rates, and usage behavior. Occupancy sensing and space utilization information offer the ability to forecast space requirements.

Space planning and utilization can be optimized using real-time IoT-driven heat maps and employee work patterns. Underutilized space can be cut back while high-use space can be better equipped and expanded. Individual space versus open space requirements can be optimized using the data collected, and room capacity can be tracked for better utilization. Space planning and reconfiguration information can be used for changes in lease contracts and flexible seating. Space management software can help determine which employees need to be on site and who could work remotely. Desk assignments, hoteling, or hot-desking options can be managed in real time.

To reduce wasted energy and improve energy conservation, occupancy sensing and corresponding energy usage can highlight wasted energy and sources for corrective action. Smart lighting that activates and turns off based on occupancy is one example. Another example is matching the HVAC systems with the actual time of use. If a tenant closes their office at exactly 5:00 every day, it doesn't make sense to keep the HVAC system running until 6:00. IoT occupancy sensing can provide trend data to make these determinations.

Space utilization metrics have become very agile post-pandemic, and cost per person has become a key metric. Occupancy and space utilization information can produce metrics around daily peak utilization by space and business unit, along with average peak utilization, the frequency of peaks, and target ratios, all to create an agile work environment.

Demand-based cleaning activities can be aligned with actual utilization patterns or even real-time sensing of each area to ensure proper sanitation within a building. Room reservation systems with sensors can detect unoccupied rooms and scheduled rooms that are, in fact, empty. Sensors monitoring cafeteria use can let building occupants know whether space is available.

Capturing a model of the space by digitally mapping the floor plan allows for easier accommodation and implementation of social distancing requirements. Analyzing the utilization data collected, areas for repurposing or consolidation can be quickly uncovered, especially for companies adopting a flexible workplace. Interactive indoor mapping can reduce time searching for open desks, conference rooms, or colleagues.

Summary

Your building's operations IoT-driven smart programs should streamline work order management, digitalize central asset tracking for maintenance and equipment, and facilitate automation for inspections and maintenance while simplifying field operations. HVAC, indoor air management, and energy systems are high-demand areas of IoT smart building programs that can help meet your building's efficiency and sustainability initiatives. Scalable tenant and occupant communications, coupled with simple amenity reservations and scheduling programs, will improve an occupant's quality of experience. Owners, operators, and asset managers can gain insights into a building's health and performance.

In the next chapter, we will delve into the highest priorities for building owners and occupant safety. First responders can react quicker to reduce injuries, save lives, and contain building assets if they have real-time access to a building's various systems and digital information. Communication between the first responders inside the building as well as externally could be the difference between life and death.

3

First Responders and Building Safety

Building owners and operators recognize their top priority is the safety of all those in or near their building. Healthy, safe, and secure building design and construction is a primary goal to prevent injuries and illnesses and to improve compliance with laws and regulations. Providing information about, access to, and control of a building system to first responders can improve crucial response times and save lives.

New technologies help first responders by giving them building information ahead of time. This can eliminate on-the-spot reconnaissance and decision-making, preventing a need to scramble to find people who know about a building and its systems. Crucial building information can be transferred to the first responders while they are en route and once they arrive. Staying connected during an emergency is critical, and smart buildings ensure there is the capability to communicate anywhere in the building during an event.

In this chapter, we're going to cover the following main topics:

- What communication capabilities first responders require, public safety standards and codes, and the most common public safety systems used in buildings.

- IoT sensors, actuators, controllers, smartphones, and other devices that can be used to control door access to improve health and safety. IoT technology enhancements such as facial recognition, scanners, and cameras can be used for building access and registration.

- Building security systems that rely heavily on cameras strategically placed around a building. These cameras are just another IoT device.

- Post-pandemic **Indoor Air Quality (IAQ)** which is a major concern for occupants, and IoT environmental sensors help monitor and control a building's air quality.

- The demand for greater bandwidth and real-time information, which drives the need for buildings' communications systems to be upgraded with IoT, next-generation cellular, and smart-building wireless solutions.

What first responders need

Whether responding to a fire, medical emergency, or domestic threat, first responders must be able to maintain communications throughout the building and the property. It is imperative that they can communicate reliably and clearly, either using radios or smartphones. Buildings built with low-e glass have poor public-safety signal and cellular coverage, resulting from signal attenuation.

Traditionally, first responders have relied heavily on voice communications using **Land Mobile Radios** (**LMRs**) from dispatchers and others on site. Many larger departments are beginning to deploy smartphones and tablets in addition to or to replace these LMRs. In cases where departments cannot afford this new technology, individuals use their personal smart devices. These devices are used for navigation applications, dispatch information, incident command, and **hazardous materials** (**hazmat**) information.

The challenge is to maintain communication in difficult environments and situations. Adding to this challenge is that these solutions must work for different responders who enter the building using different frequencies and bands. A local police department may be operating on an analog VHF network, while the fire department may be operating on an 800 MHz digital system. It is the responsibility of the building owner or operator to test their buildings for proper coverage and, if needed, install a separate communication system to meet requirements.

Public Safety Answering Points (**PSAPs**), better known as 911 call centers, allow first responders on site to precisely locate a 911 caller, as time is of the essence. Over two-thirds of 911 calls originate from wireless phones and are just as likely to come from indoors as outdoors. Indoor location capabilities, especially in multi-story buildings, need to be able to locate a person or device horizontally and vertically in a building, helping to reduce response times to save lives.

The **Federal Communications Commission** (**FCC**)'s *Fifth Report and Order* outlines the vertical (z-axis) location accuracy of plus or minus 3 meters above or below a device for 80% of wireless **Enhanced 911** (**E911**) calls (for devices that have z-axis capabilities). While horizontal location capabilities (y-axis) have been available for some time, this new requirement for the z-axis will be able to indicate what floor a person or device is on. There is a complicated phase-in implementation schedule available, as outlined by the **FCC**.

Public safety standards and codes

Public safety wireless communication requirements for inside buildings are heavily driven by standards and codes. First introduced in 2009, these standards and codes ensure that systems used by first responders can facilitate communications in every area of a building, including exit stairwells, elevator lobbies, command rooms, basements, thick-walled areas, shielded areas, and exit passageways.

A certificate of occupancy will not be issued to a building owner until the first responder communication system meets the requirements of the **Authority Having Jurisdiction (AHJ)** in their region. Typically, the AHJ is a fire marshal or fire chief but also could be an electrical inspector, or someone from the health or labor department.

It is not the intent of this book to cover every standard and code; rather, we will highlight three key codes and one standard to illustrate their intent. Our focus will be on the IoT technology and systems required to meet these codes. Codes are developed and enforced by the local AHJ and are updated every few years. You should become familiar with the standards and codes in your area.

The most common code is the **International Code Council's (ICC) International Fire Code (IFC)** 2021, Section 510, which explains the in-building emergency radio system requirements. Basically, all new buildings must provide radio coverage inside that matches the existing coverage levels of the exterior of the building, based on the jurisdiction for public safety communication systems. This code is updated every 3 years, with 2021 being the most recent update.

The second code used is the **National Fire Protection Association (NFPA)** requirements. This contains multiple sections related to in-building wireless communications requirements, covering two-way radio systems for firefighters, fire alarm and signaling codes, and emergency communication system requirements. Lastly, the **FCC** codes contain rules and testing certification requirements for public safety service radios.

Recently, **Underwriters Laboratory (UL)** released a new standard, *UL 2524*, designed to improve two-way emergency radio systems by making them faster and more reliable for first responders. Currently, it covers annunciators, repeaters, and battery backups; however, fiber systems and other communications systems are being added. This helps building owners and AHJs align on compliance, with more **Original Equipment Manufacturers (OEMs)** being reviewed and listed. The listed systems are certified to be code-compliant while meeting stringent testing requirements.

Mahesh Nanjakla, product manager of **Emergency Radio Communication Enhancement Systems (ERCES)**, and **Bi-Directional Amplifier (BDA)** at Honeywell International, Inc., outlines in *Figure 3.1* five crucial RF code requirements to protect first responders. These are the most recent changes that are driving new technology requirements to ensure occupant and first responder safety.

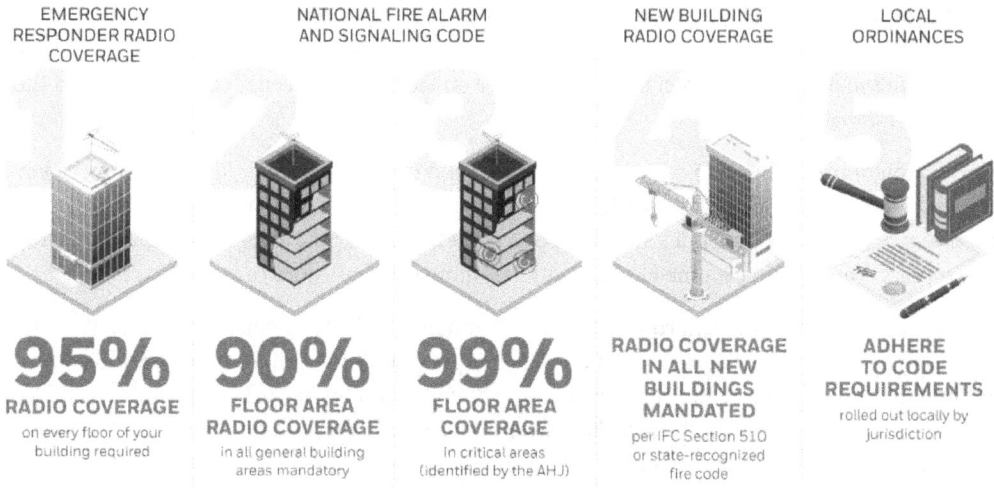

Figure 3.1 – Five crucial RF code requirements to protect first responders

The following are the most common and important requirements that must be met:

- **Wireless coverage**: 99% coverage in vitally important areas and 90% coverage everywhere else per the NFPA.

- **Equipment enclosures**: All equipment used for public safety networks must be housed in NEMA-4-compliant enclosures per NFPA and IFC requirements. These are typically made from iron or stainless steel and are watertight and dust-tight.

- **Minimum signal strength**: As per the NFPA and IFC, -95 dBm is required.

- **Antenna isolation**: The NFPA requires that antenna isolation must be 15 dB higher than the amplifier gain.

- **Battery backup**: 24-hour backup is required for public safety systems.

- **Fire rating**: All cables and equipment rooms used in the public safety system must meet a 2-hour fire rating.

- **Coverage grid**: Must meet the 20 grid or 40 grid process. Basically, each floor is divided into groups of 20 or 40 sections. Using a public safety radio, each section is tested to make sure there is connectivity and minimum signal requirements are met.

While there is agreement on the need for standards and commonality for public safety requirements in and around buildings, there remains a debate over who is responsible and, therefore, who should pay. Over the past years, we have seen this debate resolved by implementing public-private partnerships with shared responsibilities.

There is much worldwide debate about who is responsible for ensuring a building and its surroundings meet these requirements. Are the building owner and operator responsible or the local first responders' fire, **Emergency Medical Services** (**EMS**), and police departments? The answer varies by location, but as public and private communication networks evolve, we see many coming together in a public-private partnership model.

Public safety systems

Most buildings with poured-concrete structures and low-e glass create signal coverage holes and hard-to-reach places. The challenge is how to receive radio signals from outside a building without impacting communications outside the building. Typically, this will require an in-building safety communication system. These amplification systems are commonly referred to as public safety **Distributed Antenna Systems** (**DASs**), public safety **Bi-Directional Amplifiers** (**BDAs**), or public safety repeaters.

BDAs are signal boosters that boost the uplink and downlink to a remote cellular base station while filtering out any undesired signals. They do this by either passing a large band of frequencies or passing through individual channels. These signal boosters are designed for extreme conditions that usually are associated with an emergency such as fires, floods, and power outages. Their enclosures are designed to handle these conditions and can withstand high temperatures and power surges. The FCC limits the use of BDAs and requires the use of a license to operate a booster on public access channels to avoid frequency inference.

A DAS uses cables to distribute BDA signals evenly throughout a building. Multiple antennas are placed in strategic locations within a building along with radiating cables (leaky cables). These cables can be traditional coaxial cables or fiber-optic cables that provide lower signal loss and can extend over 10 miles for campus applications. A DAS combined with a BDA is considered the best option to provide capacity and coverage inside dense structures, such as medical centers, high-rise buildings, and shopping centers.

A repeater system takes the cellular signal from outside a building and amplifies the signal. Once amplified, it rebroadcasts (repeats) the signal inside the building. The system will typically have three components – an outside antenna, an amplifier, and an inside antenna. The signal is captured on the outside antenna and travels via cable to the amplifier. The signal is amplified and rebroadcast through the inside antennas.

IoT and the future of public safety systems

The United States Congress passed a law in 2012 to establish an independent authority called the **First Responder Network Authority** (**FirstNet**) and actioned it to build and operate a nationwide broadband network entirely dedicated to first responders. This network uses the latest cellular technologies to create a single platform that ensures public safety communication is always on, with priority preemption capabilities. This means that no matter how busy a commercial cellular network is, voice and data communications for public safety will always be supported with high-speed capabilities.

It is assumed that when this network is complete, in-building public safety systems will be required to include FirstNet's frequencies.

IoT sensor solutions, including Bluetooth and artificial intelligence, are helping public safety personnel make better decisions. Audio, video, and other IoT sensory data from smart devices provide emergency managers and first responders with real-time awareness and data, used for policy decisions, training, and other improvement initiatives.

Proximity and occupancy IoT sensors can help to determine the locations of individuals and first responders, and way-finding applications can guide first responders in and occupants out. Field communications capabilities are enhanced with smartphones, laptops, and tablets, as well as the use of facial recognition and fingerprint sensors. These smart public safety technologies create transparency, improve communications, and help building occupants be safer.

IoT-enabled devices play a key role in providing first responders with real-time information. For example, smoke detection sensors can trigger an immediate notification alert, both within the building and to external public safety personnel. Occupancy sensors, cameras, and noise sensors can provide insight as to how many people may be impacted. IAQ sensors can provide data about any dangerous chemicals and overall air quality, while giving first responders direct access to control the amount of outdoor air that is brought in while they are en route.

Victims' devices, such as smartwatches, health monitors, and activity trackers, can provide vital statistics for situational awareness prior to a first responder arriving on a scene. First responders could use **enhanced Multimedia Priority Services** (**eMPS**) and **Mission-Critical IoT** (**MC IoT**) devices to securely access **Electronic Patient Care Records** (**EPCRs**). Vital signs can be transmitted directly from the site and communicated to the hospital for guidance. The hospital can prepare emergency space and services, reducing critical response times by minutes.

Firefighters can access digital building information from their vehicles to review floor plans, utility layouts, and power systems to identify areas with hazardous materials, and review entrance and exit points to determine the best access method. IoT devices such as beyond-the-line-of-sight drones, robots, unmanned vehicles, and body cameras can provide faster response time and better situational awareness. With ubiquitous network connectivity, real-time data, and direct control of autonomous systems, there are many opportunities with IoT applications.

IoT and first responder challenges

Smart building technologies can sometimes create new challenges for first responders, especially when traditional mechanical functions of a building become computerized. An example is a building's elevators. In a traditional building, first responders would have access to an elevator key for manual access and control. In a smart building, a building technician would need to be called in to override the elevator system for access.

Much has been said about the ability to give first responders access to smart HVAC controls, but first responders are more likely to turn off the power to an entire building rather than try to use automated controls to turn off one area of the building. Lack of training on a building's smart systems and building owners withholding information due to privacy concerns are some other potential risk areas that need to be overcome.

Finally, the perceived reliance on an additional piece of equipment such as a smartphone can be problematic for first responders. There is additional time and effort required to bring a device online and maintain it. They also need to have confidence that the equipment will work from the start and throughout the period for which it is required.

New safety requirements for the new normal

The pandemic, violent protests, and an increase in physical assaults are driving an increase in building access control systems in commercial office towers, schools and multifamily and residential environments. These systems are designed to control who has access to buildings to ensure the safety of the assets, occupants, and more. Safety may be defined in terms of the physical safety of a person or asset, or in terms of health by keeping viruses and diseases out of a building. These systems monitor traffic, improve traffic flow, prevent crimes, reduce fire risk, improve convenience, and can be integrated with other property technology. They can even lead to lower insurance premiums.

There are several governmental egress requirements and codes, both at the national and local levels. You are encouraged to understand the requirements for your particular jurisdiction before undertaking any project dealing with building access. Fire safety requirements will vary between residential and commercial applications as well. Typically, an inspection will be conducted by a local authority before the building is occupied and then at regular intervals thereafter.

Building access control systems

Before deciding which IoT smart applications to add to your building access control system, let's review the most common types of access control used today:

- **Manual access control**: Access is granted by a human who is monitoring the access point. This could be a security guard, doorman, or receptionist checking IDs.

- **Mechanical access control**: These are traditional locks using keys or numerical tumblers.

- **Mechatronic access control**: This system uses a combination of mechanical and electrical controls, such as an **Radio-Frequency Identification (RFID)** card, key card, or key fob before using a key.

- **Physical access systems**: This system uses obstacles such as gates, turnstiles, and other barriers to manage access.

New access control technologies that include IoT and smart building system technologies include the following:

- **Biometric technologies**: This fast-growing IoT sensor or camera technology uses unique physiological characteristics from a person to develop a positive identification. Fingerprint reading is a common method but has fallen out of favor as post-pandemic no-touch requirements have been implemented. Iris scanner technology is the latest biometric technology on the rise.

- **Facial recognition**: This IoT sensor or camera technology looks at different characteristics of a person's face, such as shape, size, and the position of the nose, eyes, and cheekbones. Skin texture, heat patterns, and vascular patterns can also be used to convert data to match a known face in a database. Facial recognition is touchless and frictionless, eliminating the need for cards, fobs, and keys.

- **Mobile credentials**: This user-friendly system turns a person's smartphone, wearable gadget (smartwatch), or tablet into an access card. When coupled with a cloud-based system, credentials can be issued or revoked easily.

- **Cloud-based/hosted**: Access is provided through a web-based application using secure credentials. These highly intelligent applications are very customizable. Some offer hosted access to multiple network facilities.

Using these newer access control technologies allows for seamless integration with other business and security systems, such as the following:

- **Guideline adherence systems**: These are screening-type systems that use IoT sensors and scanners to detect elevated human temperatures, social distancing, and mask requirements, while also being capable of detecting unauthorized person(s) who may be wearing a mask to hide their identification.

- **Visitor management system**: These computer-based or cloud-based systems can handle large volumes of visitors, such as in a hospital. The system can easily be integrated with building automation systems that can handle real-time situations, including locking down areas if needed. Automated screening processes ensure compliance.

- **Video surveillance**: These systems can be integrated with HR systems and are used in large buildings and campus environments.

There are several types of access control systems:

- **On-premises access control**: The system controller or server is located within a building to increase security and improve response times. These are best suited for high-security applications.

- **Cloud-based access control**: These systems allow you to monitor and control access remotely through an internet-connected device such as a smartphone.

- **Edge-enabled access control**: These systems are very similar to cloud-based access control, except there is a storage card installed in the access controller and data is saved in the controller itself. Users access control through a smart device or computer.

- **Standalone access control**: The electronic door is connected directly to a standalone access control keypad. Programming is completed at the keypad level, and there is usually no audit trail.

Building owners and operators should consider installing encrypted technologies to stay on top of vulnerabilities. The data collected from these newer technologies can also be analyzed to measure the true hybrid/in-office ratios and support space optimization and space planning requirements. Data can be used to improve business efficiency and demonstrate compliance with public health requirements.

Visitor registration

Knowing who is in a building is essential to cultivating health and safety. Recent pandemic requirements have shifted the traditional paper-and-pen registration at the reception desk process to a safer, more distant process using the latest technologies. The basic requirement for these systems is that they must be smooth and quick while making sure all required data is captured. These systems are very helpful in office buildings, hotels, schools, and many other places.

These systems not only automate and digitize the visitor registration process but can also do the following:

- **Capture visitor information**: Not only are traditional name, number, company affiliation, and contact information collected but temperature checks, health screening questions, check-in time, and other relevant information are also now captured.

- **Notify associates with real-time alert notifications**: Personnel can receive an immediate notification of a guest's arrival using phone, text, email, or other integrations, such as MS Teams, Slack, or Google Chat.

- **Take and retain guest photos and Non-Disclosure Agreements (NDAs)**: Visitor information is captured and retained for future visits. Relevant NDA and other legal documents can be stored with the data.

- **Integrate with other software systems, log visits, and analyze data**: Integration with other business systems drives automated syncing. CRM updates can be automated and analyzed.

- **Print visitor badges**: A printed identity badge can disclose vital visitor information and a photo.

- **Develop evacuation list**: A visitor list can be emailed or printed during an evacuation to check whether anyone is missing.

Visitor Management Systems (VMS) not only improve safety and increase the well-being of building occupants but they also build brand awareness by leaving a visitor with a positive impression of the building and its use of technology. Another area that can contribute to positive impressions while improving wellness is the building's cleaning protocols.

Cleaning

Improving wellness and focusing on building health protocols have been major areas of focus post-pandemic. Many new IoT solutions were introduced during the pandemic to address the immediate need to significantly increase cleaning while minimizing the environmental impact and costs. With variable occupancy rates expected to be the norm, building operators are looking for on-demand methods of running a building, and this includes cleaning when occupancy rates dictate it:

- **Occupancy monitoring**: Occupancy sensors such as motion sensors, temperature sensors, and lighting sensors can track how spaces are used and provide insight into the cleaning requirements.

- **Restroom monitoring**: These IoT sensors also can determine when restrooms are full and provide wait-time information to occupants via smartphones and real-time notification to cleaning crews when the restroom needs to be cleaned. *Chapter 7, Smart Building Architecture and Use Cases*, provides a detailed use case based on Kimberly-Clark's smart restroom solutions.

- **Restocking**: IoT sensors are being used to sense critical supplies such as paper products, soap, sanitizers, and office space usage. They can even provide notifications when trash bins are full. Why clean a space that has not been used?

- **Tracking and compliance**: RFID tags and beacons can document cleaning activities and times automatically. This is important for building operators to demonstrate compliance with hygiene, safety standards, and certification requirements.

- **Integration and analytics**: Building operators can connect their data directly with suppliers for a more transparent, on-demand ordering process. This data will also improve forecasting, staffing, scheduling, and budgeting activities.

- **Smart cleaning digital twins**: This creates a digital representation of the spaces and cleaning requirements, and building operators can run a what-if analysis to study different cleaning plans. *Chapter 8, Digital Twins – a Virtual Representation*, explores in detail how digital twins work and the benefits that can be gained through their use.

Numerous other smart cleaning solutions are being developed and introduced by companies such as Kimberly-Clark, who deployed an entire solution suite called Onvation, and TRAX, who released their TRAX SmartRestroom.

Touchless controls

The demand for touchless technology has exploded post-pandemic as building owners and operators rethink interactions for safety concerns. Touchless technology is not new, as lights that turn on when someone enters a room, hand dryers, toilets, faucets, and automatic doors have been around for decades. This technology is defined as any device that can be used without physical touch, and the outcome goal is improved health. Apple Siri, Amazon Alexa, and Google Home voice-activated systems are common and everywhere these days.

The following are some of the most common technologies used today for touchless technology:

- **Voice assistant/activation**: Conversational or speech-enabled recognition software will perform a task based on voice commands. These can be unreliable in noisy environments.

- **Gesture-controlled technology**: This uses computer vision, radar, and AI to interpret gestures without touching the device.

- **RFID/NFC**: **Radio-Frequency Identification (RFID)** and **Near-Field Communication (NFC)** use radio frequencies to communicate between devices.

- **Touchless sensing**: Sensors detect the presence of motion, such as gesture recognition technology.

- **Bluetooth**: Low-energy encrypted signals are picked up by a reader, and then it connects a mobile app to sensors and systems.

- **Barcodes and Quick Response (QR) codes**: QR codes are essentially barcodes that provide information when scanned with a smartphone.

- **Biometric authentication**: Facial recognition characteristics are used to identify a person.

- **Hologram**: Floating images of control systems are displayed which may be controlled by the user using the gesture user interface.

Touchless technology is energy-efficient, since it automatically turns equipment/systems on and off when used. It will reduce costs, since surfaces will not need to be cleaned or maintained. Productivity and, therefore, earnings can be increased through reduced employee sick time. The overall occupant experience is more satisfying and convenient.

To reduce people's anxiety, healthcare settings such as hospitals, emergency rooms, and clinics are implementing voice-activated technology, touchless displays, and touchless kiosks. Libraries, hotels, theaters, and restaurants can have enhanced digital signage. Intelligent lockers, feedback forms, and menus use QR codes. Touchless technology will continue to grow because it offers improved safety and convenience.

IoT smart building safety systems

Many people may not realize that a lot of the security, cameras, and accessibility systems deployed in buildings today already use IoT technology. Continued development of newer solutions helps smart buildings not only expand their security and accessibility capabilities within their premises but also integrate directly with smart, city-wide security programs.

Security and cameras

Earlier in this chapter, we covered building access control systems and visitor registration systems, two highly important parts of an overall building security system. Other security systems components can include security cameras and video surveillance inside and outside of a building. Sensors and door alarm systems can activate loud noises to deter criminal activity while it's in progress. Buildings need to have emergency and fire systems to meet code requirements. Commercial cybersecurity systems protect buildings' networks and typically include antivirus software, traffic monitoring, data encryption, and firewall protection.

IoT-driven smart building solutions and management systems not only deliver lower energy costs and improve operational efficiency but also enhance building security, such as in the following ways:

- **Real-time security alerts**: Sensors connected to surveillance cameras and integrated with an access control system can facilitate the issuance of security alerts to security personnel when a breach occurs. Immediate steps can then be taken or additional authorities are called in to assist.

- **Geofencing**: Mobile applications use geographical fences or barriers to select and define very specific areas. The facilities' door locks, surveillance cameras, and other security sensors and devices can be connected and share information to create a geofence barrier around an area or entire building.

- **Security audit trails**: IoT sensors and devices collect useful data as a breach occurs, creating a historical log for authorities and insurance carriers to analyze. Building operators can use this data to enhance security procedures to prevent future breaches and possibly lower insurance premiums.

- **Connected security**: Exterior cameras and IoT sensors can provide real-time data and insight into city systems and departments, such as the police and fire departments.

- **Unified view**: When integrated with a building management system and other systems such as asset tracking, a single unified view of all a building's assets and systems can be monitored, tracked, controlled, and managed together.

Building occupants have become more aware and concerned about the safety, security, and health aspects of the building they live or work in since the pandemic. Building owners and operators have aligned their priorities on prevention, access to real-time information, and fast reliable communications. Prevention efforts are focused on who and what enters a building with security access systems, visitor registration systems, and surveillance systems adding IoT sensors, actuators, controllers, cameras, and smart building technologies to connect these systems.

Accessibility

Nearly 20% of Americans have a disability that impacts their ability to see, hear, walk, climb stairs, or grasp an object. Local and state governments have developed accessibility standards to ensure barrier-free access to and within public spaces, buildings, and government facilities. Removing barriers, developing common accessibility features such as ramps, and adding IoT smart building technology can create user-friendly access and spaces.

IoT devices enhance the quality of life for those with disabilities. Digital assistant devices such as Amazon's Alexa, smart lights, smart appliances, smart speakers, and other devices allow you to use voice-activated commands. Temperature and indoor air quality monitors, smartwatches, healthcare devices, and motion detectors can measure an occupant's health.

Location-based technologies and beacons can assist the visually impaired to navigate a building using their smartphone to provide voice directions. Similarly, way-finding applications and solutions can help people navigate in large complex buildings. Smart devices and wearables can assist those with hearing aids by providing audio effects.

Rapidly advancing IoT technology implemented throughout a building benefits those with disabilities by reading the surroundings, providing them with access to information, and performing functions for them. This helps to bridge the disability gap and creates a more inclusive and safer environment.

Indoor air quality monitoring

The pandemic has drawn new attention to the health and safety impacts of indoor air quality. Occupants want to know that indoor air is clean and safe to reduce and prevent the spread of viruses. The return to buildings by workers will largely depend on the ability to mitigate the air quality and the willingness to communicate these actions and results to all stakeholders. The **Environmental Protection Agency (EPA)** reports that individuals spend nearly 90% of their time indoors and that some pollutants are 2 to 5 times higher than outdoor levels and continue to rise with new construction materials and methods.

IAQ includes the air within and around the building and is typically tied to the health and comfort of the occupant. Concern for indoor air began long before the recent pandemic, with the **World Health Organization (WHO)** reporting that up to 30% of all buildings experienced poor IAQ in 1984. This was commonly referred to as **Sick Building Syndrome (SBS)**, a condition whereby people became sick or existing illnesses were directly impacted and made worse by the poor quality of the indoor air. Most often, the building's HVAC systems were considered the cause.

Health effects can occur immediately upon or shortly after exposure, while other effects could show up years later.

Central nervous system symptoms:	Mucous membrane irritation:
• Headaches • Fatigue • Difficulty in concentrating • Lethargy • Skin itching and irritation • Diarrhea	• Itching and inflammation of the eyes, nose, and throat • Chest tightness and asthma-like symptoms (without wheezing) • Sinus congestion • Sneezing • Nasal congestion

Table 3.1 – IAQ-related health effects

Air exchange rates, weather conditions, the outdoor climate, and several pollutant factors affect the quality of indoor air. Most contributing pollutants come from sources within a building; however, others can originate outside the building. Common indoor sources are building materials, cleaning supplies, paint, heating, and cooking appliances. Outdoor sources enter the building through doors, windows, cracks, and the ventilation system. People entering a building can also bring in soil and dust on their clothes and shoes.

Radon gases form in the ground and can enter through gaps and cracks. It is a carcinogen and the second leading cause of lung cancer. Legionnaires' disease is caused by poorly maintained HVAC systems. Carbon monoxide from vehicles on the street can enter a building every time a door is opened. *Figure 3.2* indicates the many pollutant sources contributing to indoor air composition.

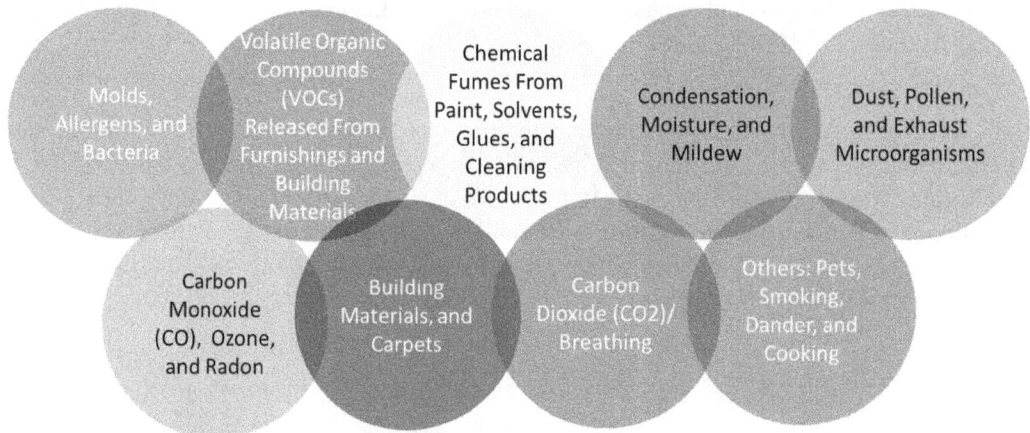

Figure 3.2 – Sources of indoor pollutants

Nearly all industry associations, government agencies, and world health organizations have made the same suggestion for building owners and operators on how to respond to the pandemic and indoor quality. The following are the recommendations, and we have indicated where IoT smart building solutions can be implemented to assist:

- **Create an IAQ management plan**: An IAQ management plan has to be created for a building and communicated to all stakeholders. The plan should outline what mitigation initiatives and changes have been made to improve the quality of the air:

    - Plan details, inspection reports, checklists, logs, and assessments can be displayed via digital signage, sent to emails, or via text using the building's IoT network

    - IoT sensors mounted throughout the building can provide real-time results for key monitored environmental conditions

- **Follow all public health guidelines**: These include social distancing, masks, personal protective equipment, occupancy levels, hygiene, sanitation, and cleaning guidelines:

    - As discussed earlier in this chapter, restroom IoT sensors can notify personnel of when cleaning has been completed or is required, occupancy levels, and supply requirements

    - IoT occupancy sensors, cameras, and heat sensors can ensure guidelines are being met

- **Improve ventilation, filtration, air cleaning, and air distribution**: Changing HVAC, **Air Handling Unit (AHU)**, and **Rooftop Unit (RTU)** filters and accurately managing the amount of fresh air are key. Promote the mixing of air without causing strong air currents:

    - Change to air filters that are capable of collecting more and smaller airborne particles. Filter performance levels are measured using **Minimum Efficiency Reporting Values (MERV)** ratings on a scale of 1 to 16, with 16 capable of reducing up to 95% of the particulates. **High-Efficiency Particulate Air (HEPA)** filters can achieve 99.7% efficiency levels.

    - IoT sensors are now being placed directly on air filters to indicate when cleaning needs to be performed or the filter needs to be replaced. IoT sensors can measure humidity, temperature, **carbon dioxide (CO_2)**, **Total Volatile Organic Compounds (TVOCs)**, **carbon monoxide (CO)**, and **Particulate Matter (PM)** levels to identify when and how much fresh air is required.

- **HVAC optimization**: Maintain temperature, humidity, fresh air supply, and flush spaces between occupied periods:

    - IoT sensors support remote monitoring and control

    - There is energy efficiency with real-time data from an IoT system

- **System verification via IAQ monitoring**: Verify that systems and mitigation efforts are functioning, as designed by continuously monitoring air quality:

    - IoT sensors mounted throughout a building can provide real-time results for key monitored environmental conditions

IoT sensors used to measure IAQ typically measure five environmental factors:

- **Temperature**: Cold temperatures can suppress the immune system and aid the spread of viruses. Temperature is measured and controlled to maximize occupant comfort and productivity. The recommended optimal range is 18–25°C (64–77°F).

- **Humidity**: This impacts comfort, respiratory health, productivity, and those with allergies, asthma, or other respiratory illnesses. Maintaining humidity can minimize the growth and spread of mold, viruses, and bacteria. Humidity levels between 40–50% are considered optimal.

- **CO2**: High CO2 levels can lead to difficulty concentrating, fatigue, and decreased cognitive ability. Concentrations below 600 ppm are considered ideal for a healthy and productive workspace.

- **PM2.5 particulate matter**: Dust triggers allergies and asthma attacks and causes eczema flare-ups. Chronic and acute bronchitis symptoms may worsen. The recommendation is to maintain dust levels below 15 micrograms per cubic meter (μg/m3).

- **TVOCs**: These are chemicals commonly found in the buildings that can cause headaches, fatigue, allergic skin reactions, eye, and throat irritation. These can affect comfort, concentration, and productivity. The recommendation is to maintain levels under 333 parts per billion (ppb).

Good ventilation, air exchange, and IAQ can reduce exposure to airborne viruses, such as SARS-CoV-2, and other diseases, chemicals, and odors. Improving IAQ makes a building healthier and can increase its value.

Building communications network/broadcast/push messaging

Building occupants always expect Wi-Fi connectivity and cellular coverage throughout a building. They are not bound to their desks and offices and roam from space to space. Applications such as indoor navigation, coupled with teleconferencing, augmented/virtual reality, and data networks drive massive increases in bandwidth requirements. Low-powered IoT connectivity networks are needed to connect the explosion of sensors, actuators, controllers, and devices being added to buildings. These wired, wireless, and IoT/data networks must be integrated to deliver a converged network infrastructure in a smart building.

Many buildings have some sort of owner-installed, owner-controlled network in place, which traditionally is a wired broadband communication network with switches and electronics. Using wires, these networks connect critical yet siloed HVAC and security, fire, alarm, elevator, CCTV, and other systems. These buildings have traditionally insisted that the tenant supplies, builds, and maintains their own enterprise communication and data networks.

The problem with this model is that smart building applications and services cannot rely on these tenant-enterprise networks to deliver the building owner/operator's smart building experience. Tenant networks vary widely in the level of sophistication, maintenance, support, and security, and tenant change is very disruptive. The trend is for building owners to deploy their own building-wide converged communication and data network. These networks will be the backbone of the building's operations, connecting its systems while providing tenant-facing services.

There are several converged network development options:

- **Local Area Network (LAN)**: This traditional network is built for voice, data, and video transmission and is sometimes referred to as an enterprise network. It uses electronic and wire cabling components in a vertically and horizontally connected network. Many buildings still have twisted-pair telecom wiring and/or Ethernet cables connecting components to servers and gateways. Surging bandwidth demand is driving the need for these networks to be upgraded to fiber or replaced with cellular or other wireless technologies. Cloud computing and data storage can eliminate the space required for servers.

- **Optical Transport Network (OTN)**: This network covers an entire building and is typically constructed by running fiber cable vertically to each floor's telecom closet and then horizontally across the floors. A central meet-me room connects the incoming service providers with the in-building fiber network. IP-protocol Wi-Fi, Ethernet, cellular small cells, DAS, and other transport options can then be connected.

- **Private cellular 5G and LTE networks**: These enterprise-located networks are compatible with public cellular networks and have the same security protocols. Who owns and operates the network can vary and is usually tied to the owner of the radio frequency license. In some cases, the public carrier will own it; in other cases, the building or business may own it, or they may share ownership in a hybrid model:

  - **Mobile Network Operators (MNOs)** place edge devices in the building and transmit them to the closest cellular tower. The tenant, business, or occupant will arrange services directly with the MNO.

  - **Neutral hosts** are like MNOs, and the facility owner will typically pay the neutral host provider for improved connectivity and services.

  - **Private enterprises** can own and operate their own LTE or 5G network to use for their facility and provide services to tenants. The enterprise would need to acquire wireless spectrum from the government or a third-party provider.

- **Non-cellular wireless**: There are dozens of different wireless technologies available for in-building communications and data systems, and they are too numerous to mention here. The more popular technologies are as follows:

  - **Z-Wave** operates in the 908.42-MHz channel and is typically used for locks, lights, and thermostat applications. It's limited to 232 devices and transfer rates of 100 **kilobits per second (kb/s)**.

- **Zigbee** is similar to Z-Wave but has transfer rates of 250 kb/s, and Zigbee operates in the 2.4 GHz frequency band.

- **Wi-Fi** is the most common technology used with transmission rates of 300 megabytes per second, and next-generation Wi-Fi is 1 **gigabit per second (gb/s)**.

- **Bluetooth** is a low-band, low-energy solution with transmission rates of 25 megabytes per second. It's one-directional and has short ranges.

- **EnOcean** has transfer rates of 125 kb/s in a low-frequency range. It supports longer distances.

- **Power over Ethernet** (PoE): In addition to providing a communication network, PoE networks can transmit electrical power. *Chapter 7, Smart Building Architecture and Use Cases*, provides a detailed use case based on the many different applications of PoE in a hotel environment.

Reaching staff, employees, teachers, and occupants during an emergency event is critical. Traditional methods such as outdoor sirens, emails, and phone calls may at times not work due to the event itself, poor coverage, or people not checking their devices, or it may not be appropriate to use given the situation (e.g., an active shooter).

Many buildings have installed internal broadcast systems that allow first responders to broadcast directly to a specific space or an entire area to reach targeted audiences within a building or throughout a campus environment. Other systems sound audible alarms, display visual information on digital signage, and/or flash the lights.

Some may use text message lists, location-based broadcast text messaging, computer pop-up notices, emails, or traditional broadcast news. There are software solutions that allow individuals to connect with everyone on a defined team and first responders within seconds. Other applications enable building operators to digitally manage their building's emergency plan, drawings, procedures, and compliance practices via a browser or smart device.

Summary

Having the capability for first responders to communicate in a building is critical for occupants and the responders themselves. Many standards and common IoT technology solutions have been implemented to ensure uniformity across localities. A new normal post-pandemic building environment is driving new requirements to build access, touchpoints, IAQ, and cleaning solutions, all of which are being solved with IoT smart building technologies.

The next chapter will explain how to enrich building data with location context to make a building smart. Space planning, indoor mapping, people counting and tracking, occupant engagement, asset tracking, and room and desk booking all benefit from location-based information.

How to Make Buildings Smarter with Smart Location

Smart buildings are seeing an increased focus on occupant-centric workplaces using occupancy analytics and **Real-Time Location Systems (RTLS)**. RTLS generate live updates on the location of people and objects. Think of them as the blue dot on a map showing where something is at a given moment.

Memoori defines occupancy analytics and location-based services in the commercial office space as "*electronic hardware and software used in workplace management and indoor location-based services to enable one or more of the following applications – optimizing spatial efficiency in the workplace, improving the location-based experience of building users, people finding, using location-based data to optimize building and operational performance, asset tracking.*" Memoori estimates that the occupancy analytics commercial office space industry will grow at a CAGR rate of 21.5%, from $2.17 billion in 2019 to $5.73 billion in 2024.

Location-based sensors and devices are being used to increase efficiency, improve safety, and enable a more enhanced user experience. Smart locations are physical locations equipped with networked IoT sensors to give owners, occupants, and managers more information about the condition of those locations and how they are used. Space utilization analytics can help strategic planning for new spaces, growth needs, or consolidation requirements.

COVID-19 has increased the focus on RTLS combined with IoT sensors, analytics, and automation to monitor and manage building conditions and processes. Space planning, indoor mapping, people counting and tracking, occupant engagement, asset tracking, and room and desk booking all are driven by RTLS and IoT technology.

In this chapter, we're going to cover the following main topics:

- Collecting location data through thousands of IoT sensors, which is fundamental in delivering occupancy analytics and location-based services.

- Locating people or objects using an indoor positioning system, comprised of a network of IoT devices to cover indoor locations where GPS and other satellite technologies do not reach.

- Navigating through smart buildings using off-the-shelf solutions and IoT sensors, beacons, and devices. You can precisely locate people and assets within any location in a building by just using your smartphone, smart device, or tracking tags

- Discovering how real-time occupancy analytics using IoT sensors improves space optimization, people counting and tracking, occupant engagement, room and desk booking, cleaning, and much more.

While some of the technology outlined in this chapter may seem to be repetitive or have similar terminology (that is, **Wireless Internet for Frequent Interface (Wi-Fi)** and **Bluetooth Low-Energy (BLE)**), it should be noted that while the technology is similar, the application and method of use are often different, and special care should be taken to note the differences.

Location data sources

RTLS are all about collecting and analyzing data. Therefore, it is important to understand where that data is coming from and what its source is. Knowing how the data is collected is extremely important, as it will help you to determine the data's accuracy and depth of information. Location data is information collected at the device's geographical position, called coordinates, and is typically expressed in the latitude and longitude format.

There are several ways in which location data is collected:

- **Global Positioning System (GPS)**: Latitude and longitude coordinates are collected on a device by communicating directly with satellites. The location is calculated by measuring the time it takes the signals to be received. Mobile phones, car navigation systems, and fitness trackers typically use GPS. While GPS is considered to be the standard for location data because it is very accurate and precise outdoors, it is difficult or impossible to use indoors because of physical obstructions created by a building.

- **Software Development Kit (SDK)**: Software app developers place SDK code in their applications. This location code is collected directly from a device and typically requires a user's permission, or it can run in the background. The data is very accurate and trend data can easily be compiled to show daily habits. One challenge, however, is that it requires user permission to opt in, making it difficult to use at scale.

- **Bidstream**: This data is collected from ad servers when an ad is connected to a website or mobile app. While easy to collect, this data is usually inaccurate, incomplete, or not legitimate. For example, it could collect an **Internet Protocol (IP)** address that is routed through a **Virtual Private Network (VPN)**, creating a completely different location.

- **Wi-Fi**: Wi-Fi devices emit probes to locate a router, and these probes can be measured to determine the distance between a device and a router. How well the Wi-Fi network is built will determine its accuracy and precision. This method works better than cellular and GPS indoors when those signals are obstructed.

- **Beacons**: These are hardware transmitters that, when in proximity, can sense other devices. While the location data is very accurate, these transmitters would need to be placed in massive amounts to scale effectively.

- **Point of Sale (POS)**: This data is collected from a transaction with a customer, such as a cash register in a building's store or cafeteria. While large amounts of additional data can be collected, it will only collect data when a transaction occurs and may not transfer all the data if the transaction is quick. POS systems tend to be decentralized, making trending difficult.

- **Point of Interest (POI)**: This data describes a building's location and is used with other data types to develop insights about traffic and behavior.

Location data will typically have attribute data fields beyond latitude and longitude. Horizontal accuracy, altitude, and elevation information usually accompany latitude and longitude. Timestamps can log a single event or a sequence of events.

An **IP address** is the number assigned to every device connected to a computer network and is used for location addressing and network interface identification. Smartphones each have a unique 36-character identifier called a mobile ad ID or device ID, and as you might expect, iOS and Android devices use different identifiers.

Indoor positioning

Indoor Positioning Systems (IPS) refer to the technology and systems used to locate people and objects indoors. These systems are the backbone to support location-based indoor tracking systems such as way-finding, inventory management, people locators, and first responder location systems. There are several different technologies that can be implemented, and these are the most common:

- **Proximity-based systems**: These systems detect the general location of a person or object by using tags and beacons. These are generally low-cost systems and are used in manufacturing, industrial, and healthcare-type facilities:

  - They can deploy reader-based dumb-tags that transmit their identification information continuously to reader devices, and based on the signal strength, the position can be collected and calculated.

  - Another method is a reference-point-based system that uses low-energy beacons (BLE) as reference points. The tags use the reference points to calculate their own location and communicate with a server. The access points are placed about every 100 feet apart and have a prolonged battery life.

- **Wi-Fi-based systems**: Wi-Fi transmitters send simple packets or tags to Wi-Fi access points. The access point timestamps the data packets and determines the single strength. Specific algorithms then use this data to calculate detailed positions. These systems generally have an accuracy of within 3 to 5 meters. Wi-Fi tags are costly, at $40.00 to $60.00 per tag, but they are very power-efficient.

- **Ultra-Wideband (UWB) systems**: A UWB system uses three or more ultra-wide-band readers to send a wide-range pulse over the **gigahertz (GHz)** spectrum. These readers listen for chirps that are being sent from ultra-wide-band tags. The pulse is generated on an exciter that has a spark-gap style, generating short, coded, very wide, and instantaneous bursts. These readers are very accurate but tend to be expensive, since the limited range of tags requires large numbers.

- **Acoustic systems**: This newer technology uses ultrasonic pulses from tags and is very similar to UWB systems. These systems use sound humans cannot hear instead of UWB radio signals. There are one or more receivers that pick up the sounds and locate the tags. A sonar-based system can be just as accurate as UWB systems but very costly. These systems are not widely used yet, and we believe that healthcare facilities will begin to deploy these.

- **Infrared (IR) systems**: These systems use infrared light pulses (similar to a TV remote) to locate signals in a building. IR receivers need to be installed in every room that would need to be read by a receiver device. Light cannot travel through walls, making these very accurate systems. The tags are low-cost and long-lasting, but the system does require a wired IR reader in every room.

- **Radio-Frequency Identification (RFID) systems**: RFID systems use a low-powered radio frequency to send and receive messages from asset tracking tags. Passive RFID systems send out a radio signal that wakes the tag, similar to the way toll booths work. Active RFID uses battery-powered sensors to connect with access points.

With a number of different technology solutions available, IPS is quickly gaining popularity in hospitals, airports, shopping malls, manufacturing facilities, and other indoor venues where location-based and navigation services are valuable. In the next section, we will review some of the applications that can be used with the technologies discussed in this section.

Navigation, maps, and way-finders

Indoor navigation systems are digital processes that help individuals find their way around indoor areas. These systems are very precise and leverage **Augmented Reality (AR)** to deliver precise real-world directions. Building technicians can use a navigation system to find maintenance access points. Visitors can be guided to key locations without the expense of an escort.

In hospitals, patients and visitors can be guided to treatment rooms, cafeterias, stores, or even other buildings. First responders and rapid response teams can be given immediate directions, and when connected with other systems in a building such as an elevator, its doors could be held open until they arrive. Retail stores use these systems to help customers find products. Customers can download an app and then have access to a product's shelf location, as well as the restroom and other locations.

Indoor positioning systems determine in real time where a person or object is within a given space. Navigation systems calculate the most efficient path and provide directions from a starting point to completion. Way-finders work like the navigation systems on your smartphone, providing step-by-step instructions. Asset management systems locate physical assets, as depicted in *Figure 4.1*.

Figure 4.1 – Way-finding, navigation, and real-time asset location

In its simplest form, users access a floor map via an app or browser window with their smart device. They search for a location and are then guided by interactive step-by-step navigation dots or verbal instructions. The location is continuously updated and is generally accurate to 1 to 5 meters. Common uses are museums, offices, stadiums, airports, shopping malls, and hospitals.

There are even dedicated navigation apps available to building owners and occupants/visitors. With the help of Gaurav Bidasaria at TechWiser, we have identified these as the best:

- **Microsoft's Path Guide**: This Android-only app uses sensors, a barometer, and a magnetometer in a smartphone to collect data to draw an indoor map. The app can then calculate the number of steps, turns, and levels.

- **Google Indoor Maps**: Building operators add their floor plan and layout to Indoor Maps within the Google Maps app. Visitors simply open the app and look for the building they want.

- **Anyplace Indoor Service**: This free and open source service is available on GitHub. Building operators add their floor plans, images, POI, and other buildings if it's a campus environment.

- **Situm Mapping Tool**: Situm uses Bluetooth, Wi-Fi, and smartphone sensors such as Gyroscope to develop an indoor map. The map is stored and then shared with visitors.

- **Be My Eyes**: Designed for the visually impaired, users use the app to ask for assistance. Someone with access to the app near the user will guide them to their destination. While it's not the best use of technology, it does provide navigation.

- **Indoor Radio Mapping**: This system uses Wi-Fi and Bluetooth to provide direction. Buildings would need to install iBeacons to communicate with smartphones using Wi-Fi or Bluetooth.

- **Indoor measurements**: This kind of map generator app requires beacons to be installed in a building. Floor plans and images are uploaded, and most types of beacons are supported.

- **Indoor GPS**: With this popular type of app, building operators upload floor plans and images. Instead of using Wi-Fi or Bluetooth, GPS is used.

Way-finding solutions are more engaging for users if they are intuitive, appealing, and interactive. Solutions have also been developed for sight-impaired individuals using their smartphones' speakers/headphones.

Asset location and tracking

The terms *asset tracking* and *asset management* are used interchangeably and refer to the process of keeping track of the physical items, devices, and equipment used to support the operations and maintenance of a building. Important information such as location, status, maintenance records, usage, and user manuals are often tracked along with an item. Inventory management can be performed automatically and in real time. Configuration management information can provide critical dependencies and relationship information. Assets can be monitored through their life cycle, providing building operators with useful long-term information.

The goals of an IoT-enabled asset-tracking solution are as follows:

- Real-time location to save time and money looking for assets

- Automation to eliminate manual processes and inventory management

- Increased accountability

- Improved security and loss prevention

- Maximized utilization rates

- Data collection used to make decisions

The complexity of an asset management system will be driven by the type of facility, business, and assets a building supports. Common examples include the following:

- **Hospitals**: Tracking expensive mobile medical equipment such as infusion pumps, heart monitors, wheelchairs, beds, and so on, managing facilities, equipment, and tools, inventory management, and tracking mobile computing devices

- **Commercial real estate**: Tracking furniture, electronic and IT equipment, printers, supplies, tools, and testing equipment

- **Industrial**: Tracking inventory containers, pallets, tools, indoor vehicle fleets, and equipment

- **Schools**: Tracking and locating IT items such as computers, tablets, printers, and software, sporting equipment, A/V equipment, shop equipment, and maintenance tools

Physical devices are tagged, scanned, and connected to asset management software and linked to databases. Important information about the asset can be viewed and updated using a smartphone or smart device. Where the asset is located, who is using it, who has responsibility for it, the condition of the asset, its maintenance history and schedule, and even the user manual can all be monitored and stored in one asset management system.

Asset-tracking technologies

Tracking tags located in or on an asset and the sensors used are considered IoT devices. They collect, store, and communicate the asset's information. How these devices and sensors communicate back to a central asset management software solution can be handled in several different ways:

- **Barcode tracking**: A barcode label is affixed to an asset and scanning devices are used at strategically placed choke points to read the barcode. The data is sent to a centralized computer, a cloud, or another database application. Barcodes were the earliest form of tracking and have been around for over 50 years. The barcodes themselves are inexpensive, small, and very lightweight. Most scanners have been replaced with personal smartphones that can perform scanning and data transfer actions. Barcode systems are typically the more affordable solution, since they do not require power.

  Some of the disadvantages include the inability to show real-time status and location, the requirement for line-of-sight manual scanning, being prone to human error, and the requirement for expensive software components to manage the inventory. These systems also require choke points such as doorways or checkout stations, which can be expensive to install and maintain.

  Barcoding systems are best suited for large quantity and low-value asset applications. New coding technology continues to be developed, such as matrix barcodes (QR codes) and invisible codes.

- **RFID tracking**: Low-cost electronic RFID tags are affixed to an asset and they broadcast the asset's location. The signal is automatically picked up by stationary electromagnetic readers or hand-held readers. These systems can track large amounts of information, work faster than barcoding systems, and can be used for security checks and inventory processes.

  This system does require its own network and, therefore, can be expensive. The range will depend on which frequency is being used. Key fobs use a low frequency and asset tracking uses an ultra-high frequency for greater distances. Passive RFID tags use power from the reader, and therefore, the tags do not require power, are inexpensive, and last a long time. Active RFID tags require a power source to be connected to the tag. While allowing for a greater range of up to thousands of feet, they are more expensive.

- **UWB**: UWB is fast becoming a top-performing technology that is scalable, accurate, secure, and reasonable from a cost perspective. UWB transmits over a large (wide) section of the radio spectrum by emitting short pulses. It uses a small low-cost tag that can send large amounts of data.

  UWB is best suited for heavy industry and for locations where GPS and cellular solutions are not options. High-valued assets, risky situations, and concerns for safety and security are typical applications. UWB transmissions are difficult to comprise or intercept. Because it can penetrate walls and does not require line-of-sight reading, UWB is often used in hospitals and locations where metal and reflective surfaces exist.

  UWB is relatively new, and its unique hardware is expensive and not widely available yet. It also doesn't provide true locational data. This system is not power-efficient and requires complicated, expensive, and time-consuming time-syncing laser measurements between devices.

- **Wi-Fi Positioning System (WPS)**: This system uses Wi-Fi-capable asset tags with built-in batteries placed on an asset. The WPS can be integrated into a building's existing Wi-Fi network, saving money; however the Wi-Fi-enabled tags are expensive and need more power than RFID. This system is best suited for indoor use and small numbers of assets.

  Since Wi-Fi already is well known and has wide adoption, numerous Wi-Fi-enabled devices already exist. It is suitable for localized tracking and small-scale projects. Positioning accuracy will generally be within a few meters, and scalability is difficult. It is not recommended for mobile or moving asset environments. If the asset tracking network is on the same Wi-Fi network as the internal communications, network congestion can become a major issue. Wi-Fi is not very secure; however Wi-Fi 6's new architecture does address many of the previous version's security issues.

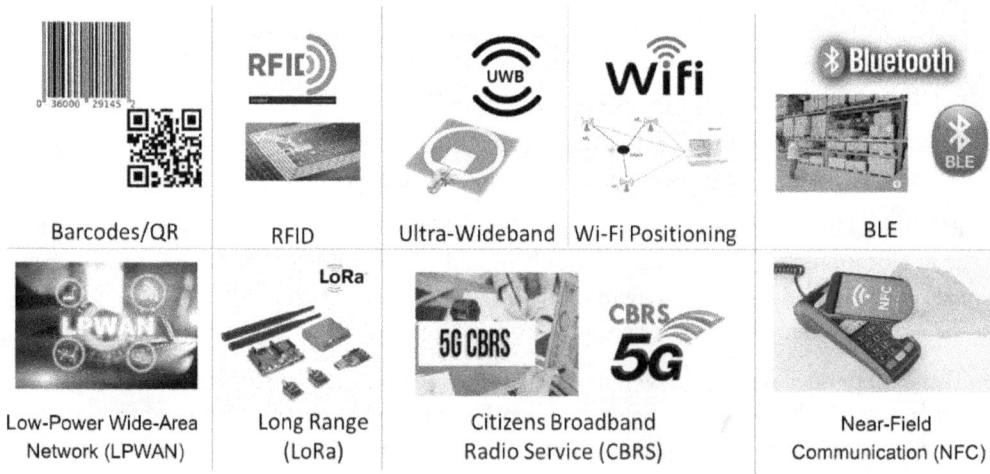

Figure 4.2 – Major asset tracking technologies

- **BLE**: BLE's low power consumption makes it one of the most common technologies used for asset tracking, and it has been available for over a decade. Often, this sort of network is referred to as a beacon network, since it is comprised of beacons and hubs. Most consumer mobile devices incorporate Bluetooth, which makes BLE a popular technology for asset tracking.

  These systems typically do not require manual scanning or searching. By using a **Received Signal Strength Indications (RSSI)** architecture, proximal locations can be determined for items that are moving. The **Angle of Arrival (AoA)** and **Angle of Departure (AoD)** are direction-finding methods used to calculate the location of an asset.

  Many of the enterprise Wi-Fi access points often already contain BLE radios, allowing an asset-tracking system to traverse on a building's IP network. BLE systems require little battery power and can cover large areas. Latency is a disadvantage, since there are many devices competing for the spectrum, and signal interference can be an issue.

- **Low-Power Wide-Area Network (LPWAN) technologies**: LPWANs have low-power consumption combined with a higher signaling range, which allows these systems to operate at long distances while consuming little power.

  Long Range (LoRa) is one of the more popular technologies in this category, and it delivers low throughput in long-distance communications. It uses lower megahertz unlicensed frequencies such as Wi-Fi and can transmit over large distances.

  Sigfox, **NarrowBand IoT (NB-IoT)**, and **Long-Term Evolution for Machines (LTE-M)** communications are other similar technologies. These are best suited for manufacturing plants and low-tech warehouse applications. They have a great range and use little battery, but on the downside, their transmit and receive data speeds are slower than other technologies.

- **Private 5G and CBRS**: **Citizens Broadband Radio Service (CBRS)** is a license-free spectrum 5G private network used in large offices, warehouses, and manufacturing campuses. It can leverage an existing IP network. This is new technology, and 5G-capable asset tags are just becoming available. These systems are great for large areas and benefit from long-lasting batteries, high-speed data, and low implementation costs.

- **Near-Field Communication (NFC) tracking**: Tags are connected to an asset, and when moved near a system monitor, communications occur. Think of the bulky plastic tag placed on expensive clothing in a store that triggers an alarm when it passes through the door.

- Other less common systems include the following:

  - **Web applications**: Hardware with an **International Mobile Equipment Identity (IEMI)** ID can be instantly traced. This is a 15-digit code in a device or piece of equipment that is identified by cellular systems. A SIM card in mobile phones and smart devices is an example.

  - **Computer vision**: Facial or object recognition is computed by a computer. These systems are highly precise and can identify QR codes and small tags. In some cases, a system can recognize items and an asset tag is not required. However, this system requires line-of-sight reading and special lighting conditions, which could be costly and complex.

- **IR**: IR systems require line-of-sight reading and provide image information only. They cannot detect location on their own but can be paired with location technologies.

- **Radar, ultrasonic, Zigbee, and others**: Radar can be used for motion detection and proximity sensing and also track people. Ultrasonic sensors generate highly pitched sounds, but these sounds can often be blocked by physical barriers. Zigbee and a long list of other technologies are becoming available, but they tend to come and go quickly due to interoperability issues.

With many choices and no one clear system a best fit for all scenarios, determining the right asset-tracking technology will require some cost/benefit analysis, based on your building type and size and the types of assets to be tracked. RFID, Wi-Fi, and BLE may be the better option for small-to-medium-size indoor applications, while LoRa, GPS, and 5G may be the better option for large, campus, outdoor, and long-range tracking needs.

Real-time occupancy

When the pandemic first began in March 2020, many grocery stores and box stores remained open because we needed food and supplies. Government and health organizations all suggested safe distancing and limiting the number of people in the store at the same time. Since most of these stores did not have occupancy sensing technology installed, they resorted to physical counting methods.

Lines developed outside stores, and using the health and safety guidelines, management determined a safe number of customers to allow in a store at the same time. Employees with cell phones were positioned by each entrance and exit, and every time a customer or customers left the store, the person at the exit door would text or call the entrance door personnel to let them know a person or several people could enter.

This accelerated the need for buildings, regardless of type, to install real-time occupancy sensing solutions for more sustainable building management procedures. Other smart building occupancy sensing applications quickly followed to create optimal working and living conditions.

Occupancy sensing applications

The previous example of a store's use of occupancy sensing is just one of many different reasons and applications. Office buildings, airports, shopping malls, and subways are just a few of the different building types implementing occupancy IoT devices and applications:

- **Health and safety**: Many of the requirements put into place during the pandemic will remain in place permanently. Occupancy sensors that can count people will assist with social distancing requirements. Some sensors can measure the distance between people to ensure compliance. Heat-sensing sensors can measure a person's temperature.

- **Space planning**: Occupancy sensors can provide the ability to monitor real-time vacancies and schedule assignments. When a desk is no longer being used, a notification can be sent to the cleaning crew to prepare the space for the next use, and space planners can schedule the next person in.

- **Space availability**: Building management and occupants can determine restrooms, exercise rooms, conference rooms, and other common areas' real-time occupancy levels to determine a room is free or not.

- **HVAC fine-tuning**: Knowing how many people are in a space by measuring occupancy or CO_2 levels can help with real-time outside air handling requirements, improving tenants and workers' comfort levels.

- **Security**: If there is an emergency, occupancy sensors can provide information about where everyone is and alert them to the nearest emergency exit, or help first responders locate individuals.

- **Energy conservation**: Motion-activated light sensors can reduce wasted energy from lights left on in unoccupied areas. Demand ventilation procedures can better control the amount of outside air that is required to meet safe air requirements while eliminating too much air that does not need to be heated or cooled.

The benefits of real-time occupancy sensing include the following:

- Compliance with government social distancing guidelines

- Staff and occupant health and safety

- Communication between occupants and improved quality of experience

- Analytics – collecting data and making informed decisions to reduce operational and energy expenses

- Optimizing space management and scheduling decisions

- Boosting productivity through collaborative informal work environments

While there are many different occupancy sensing needs and applications, there are just as many different technologies that can be implemented to deliver information.

Occupancy-sensing technologies

Smart sensors can be deployed around a building to detect an individual's presence. They then either initiate a reaction locally, such as turning on the lights, or they send real-time data to a centralized management system, which may adjust the temperature and ventilation system.

Occupancy sensors may use one or a combination of the following technologies:

- **Smart lighting sensors**: These lighting systems detect motion and turn on and off, based on the motion in a space.

- **Cameras and people counting**: Cameras are now integrated with processing functions, including people counting and face/body detection. These are very accurate; however, they tend to be expensive. In many countries, these systems raise privacy concerns and are not allowed.

- **BLE solutions**: Like the BLE solutions discussed in the asset-tracking section of this chapter, BLE is also used to track and count people. When used for people tracking, BLE tracks the number of smartphones entering, remaining in, and leaving a space. This assumes that almost everyone has a smartphone and that their Bluetooth is enabled, which often is not the case.

- **Passive IR (PIR) sensors**: These sensors process thermal signals in a surrounding space and can detect when the signals are interrupted by body heat emanating from a human. These easy-to-install, low-maintenance sensors are stuck to ceilings and walls or placed under chairs and desks.

- **IR Time-of-Flight (ToF) sensors**: ToF sensors project a beam of IR light that reflects off the individual and returns to the sensor. The time that it takes is used to determine the distance of a person. The sensor can also determine the movement direction to determine whether the person is entering or exiting the space.

- **IR array sensors**: These sensors measure a person's temperature as they move closer or further away from them. The field of view expands as the proximity to the sensor narrows. No images or personnel information is captured and stored, and therefore, they are privacy-compliant in most countries.

- **Ultrasonic sensors**: These sensors deliver high-frequency sound waves to a space to determine whether the room is in a steady state or not. If it detects disruptions of this steady state, it assumes there are people there. The sound cannot be heard by humans.

- **Microwave sensors**: These are similar to ultrasonic sensors sending high-frequency microwaves into space and determining whether there is a reflected pattern. When objects or people move, the reflections change. Microwaves can penetrate most building materials and have higher sensitivity and greater range.

- **Microphonics technology**: These sensors contain microphones to hear sounds that detect occupancy. These are useful in rooms with obstructions such as restroom stalls or office cubicles. They can detect talking, eating, keyboard clicks, and other sounds while masking building noises such as HVAC and air currents.

- **Hybrid sensors**: These are sensors that can use a combination of technologies together, such as a PIR sensor combined with an ultrasonic sensor.

Both occupancy sensors and vacancy sensors use one or more of the technologies described earlier; however, they function slightly differently. An occupancy sensor turns on lights and/or HVAC when activity is sensed, while vacancy sensors shut down lights and systems when no activity is detected. A vacancy sensor can only turn off systems and will require manual intervention to turn a system back on. Dimming sensors are a combination of features from occupancy and vacancy sensors and will dim lights according to the daylight environment.

Summary

RTLS continue to gain momentum as the explosion of IoT capabilities and post-pandemic space requirements drive the need to know where people and assets are all the time. Occupant health and safety, operations and energy efficiency, and the desire to access information instantly are fueling this growth. While GPS is the de facto choice for outdoor tracking, dozens of technologies have emerged, ranging from simple inexpensive barcoding and Wi-Fi solutions to hybrid solutions, using UWB frequencies and sounds to track and locate people and assets within a building.

In this chapter, we reviewed the navigation, indoor positioning, asset tracking, and real-time occupancy IoT technology solutions available to make your building smart. In the next chapter, we will explore another fast-growing smart segment focused on improving the quality of experience for a building's occupants. Almost any aspect of the occupant-building interface and occupant-building management interactions can be automated or made better with the addition of IoT and smart building technologies.

5

Tenant Services and Smart Building Amenities

Building operators and tenants interact regularly regarding lease agreements, monthly rent, fee payments, building operations, maintenance impacting tenants, tenant-specific requests, and numerous other building-related activities and events. Tenants' return to buildings post-pandemic is driving an increased desire for real-time communications from building operators about health and safety mitigation activities. IoT smart building systems digitize the collection and distribution of data to enhance the interaction process.

Increasing densification rates combined with inadequate space utilization are driving the need for workspace management systems. Smart buildings are seeing an increased focus on occupant-centric workplaces, using occupancy analytics and live updates on the location of people and objects. Think of it as a blue dot on a digital map showing where something is at a given moment.

This chapter will explore the numerous applications that have been developed to improve aspects of the experiences of building management and occupant interaction.

In this chapter, we're going to cover the following main topics:

- Using an example of a request for temperature change, we will demonstrate how IoT smart building solutions can automate and even eliminate the work order process.

- The difference between space occupancy, space utilization, and space use frequency, and the market drivers to add IoT solutions to improve space management.

- How IoT scanners, smart devices, and smart systems can improve the financial and informational interactions between building management and tenants.

- How IoT sensors monitor and collect massive amounts of data. Data analytics is the process of analyzing data with specific analytic tools and techniques.

- How workforce applications are adding IoT devices and smart applications to manage facilities and improve operational efficiencies.

- Building amenities and tenant services, which are adding IoT smart building solutions to improve satisfaction and create new revenue streams.

- Other applications typically not considered building-related, which can be introduced with IoT and smart buildings through a building's tenant app.

Work orders and tenant requests

In *Chapter 2, Smart Building Operations and Controls*, we briefly covered work orders and how they are at the center of any maintenance and operation system. They are a simple tool to define a task that needs to be completed, from scheduling, assigning, and tracking job tasks to completion. Digital work order systems allow work orders to be completed on a smartphone. The status of the workflow is updated at each step using IoT, barcodes, QR codes, and other smart sensors. Work orders are used to provide real-time status and prioritization, and they provide a digital record that can be used for invoicing, audits, and historical record keeping.

TENANT MAINTENANCE REQUEST FORM
TENANT INSTRUCTIONS

All general maintenance must be reported to our office in writing. In order for a repair to be attended to, please complete this form and fax, post, email or deliver to our office.

In the event of an emergency repair, contact our office immediately!

Once we have received the request, either our office or a tradesperson will contact you.

Date: _____

Time: _____

Address: _____

Concern: _____
_____
_____
_____
_____

ACCESS DETAILS
Tenant Name(s): _____

Phone Home:_____Work: _____   Mobile: _____

Access to property: ☐Take office key   ☐Tenant will be home

Tenant preferred time and date: _____   Tenant authorises entry   Yes☐

I hereby authorise your office and/or the tradespeople to enter the property with the keys in order to carry out the repair or view the repair.

Signed_____

If the repair relates to any of the following appliances, please list the make and model

Stove_____	Washing Machine_____
Oven_____	Microwave _____
Dryer_____	Fridge_____
Dishwasher_____	Air Conditioning_____
Hot Water Service_____	☐ Gas☐ Electric

Figure 5.1 – A sample tenant request form

While that simple definition focused on the building owners and operators' perspectives of the work order, this chapter will focus on the tenant's perspective. Dozens of new software applications have been introduced in just the past few years to elevate the tenant experience with work orders/service orders. These solutions allow tenants to submit requests via the web or mobile app at any time. Tenants have visibility of the work order's status and estimated completion time, and they can then provide real-time feedback and reviews.

By installing IoT sensors throughout a building, as described in *Chapter 4, How to Make Buildings Smarter with Smart Location*, worker movements can be tracked and work orders can be updated immediately. Beacons and asset tags situated all over the building establish control points that supply data on where a building engineer or repair person has been. QR codes installed in the restrooms can be scanned by the cleaning crew upon completion to date-stamp and time stamp the cleaning, while notification can be sent to management and others.

Another method is to use IoT solutions to eliminate the need for a work order altogether. If a tenant in a traditional building wanted to have the temperature changed in their work area, they would phone the building's maintenance department and open a request. The maintenance department would record the request by opening a work order, then they would locate a building engineer, and assign the task. The engineer would then stop whatever they were doing and proceed to the tenant's space to manually change the temperature.

In many cases, this process could take anywhere from 30 to 45 minutes, creating tenant frustration. Many times, the entire process occurs again when the tenant creates another request to change the temperature. The maintenance department's work order not only incurs the cost of the operation for the engineer's time to complete the task but also includes the loss of value from other work that might otherwise have been performed.

Instead, smart IoT thermostats could be installed throughout the building, and tenants given mobile device access to manage their space temperature in real time. The maintenance department could establish thresholds of a few degrees either way to control costs and keep temperatures within a reasonable range. The tenant is satisfied because they feel like they have control, and they didn't have to wait 45 minutes for the temperature change. The building operator is satisfied because they have one less work order to process, and the expense associated with each order is eliminated.

There are hundreds of different tenant requests that could be submitted. To demonstrate the potential magnitude of possibilities, the following are some examples:

- **Housekeeping requests**: This includes a janitorial service, trash removal, spill cleanup, space cleaning, table and chair setup, equipment use, sanitation, and bulk trash pickup scheduling.
- **Safety and compliance requests**: Examples include disability access equipment installation, ramps, handrail installation, smoke alarm testing, sensing water, gas or other fluid leaks, and overall general fixes that are part of the costs of running a business.

- **Tenant-specific requests**: These enhancement requests are customized for a tenant and generally will increase the rent. Examples include space buildouts, upgraded flooring, fixture upgrades, and other space improvements.

- **Property value improvement requests**: These work orders improve the value of a building and include items such as carpeting and flooring in common areas and HVAC systems in a whole building.

Smart work order solutions can streamline the request workflows to improve efficiencies and eliminate barriers. They assist in building business cases and help determine facility costs and performance, while improving contractor and staff resourcing. Pictures and other data files capture a complete recording of the work order task, and real-time access keeps everyone updated on the status while creating a historical record.

Space occupancy, utilization, and booking

Space utilization rates in 2018 were averaged 60% to 70% according to a benchmarking report by JLL. Post-pandemic rates fluctuate widely, with much lower estimates at 30% to 50% as of late summer 2022. Hybrid work policies coupled with constantly growing and shrinking tenant company requirements based on the economy are making space planning more challenging.

Traditional approaches to space utilization use rough information gathered from scheduling records, sign-in logs, and often manual observations and on-site assessments. These can be time-consuming, inflexible, not quantifiable, and outdated, missing key trends such as offices being shut down during the pandemic and making observation information outdated and no longer valid.

Space utilization is calculated as the total desks available divided by the total occupied desks, delivered in the form of a utilization rate. If there are 100 desks available and 60 desks are occupied, the utilization rate would be 60%.

Space occupancy measures the number of people in a space at the same time. Prior to COVID-19, for example, a particular office space was determined to have a maximum occupancy of 100 people; however, post-pandemic, with concerns about indoor air quality and social distancing requirements, that same space may now comfortably accommodate 65 people.

Space frequency measures the amount of time the space is used and compares it to the amount of time for which it is available.

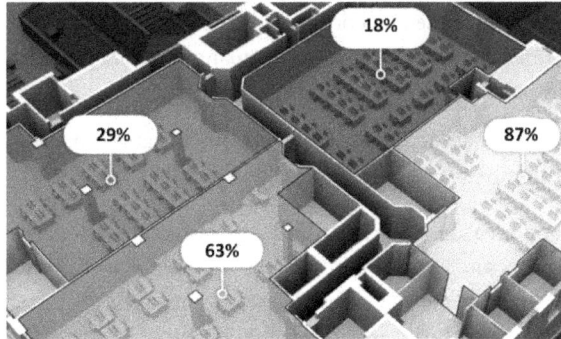

Figure 5.2 – Space utilization

Post-pandemic, many companies moved to a hybrid model of working 2 or 3 days in the office, and 2 or 3 days from home. This has driven office environments away from assigned seating and created requirements for new ways to measure space occupancy and space utilization. IoT technology and smart building solutions play an important role in measuring space occupancy in these new environments. The following are a few examples of how this technology is implemented, using the techniques discussed in previous chapters:

- **Badge sweeps**: Using ID badges with RFID tags, beacons, and sensors to track the number of people in an office space by calculating the difference between the number of people that entered and how many people exited the space.

- **Desk hoteling**: This is a space-sharing practice whereby employees, contractors, consultants, and guests are required to reserve workspaces for each day they are planning to be in the office. It is called desk hoteling, since the reservation process is like a hotel reservation. The concept is that there are fewer desks, and therefore, less space is required when it is shared. Since the worker may be at a different desk every day, additional technology solutions are often needed:

    - **Cloud document storage**: This allows a worker to work anywhere and everywhere there is an internet connection. This reduces the risks of lost data and keeps everyone up to date on the latest versions.

    - **Clean desktops**: Based on a schedule, desks can be cleaned and disinfected before the next user. With IoT sensors, cleaning crews can be notified right away when a person leaves the desk/space.

    - **Wi-Fi and cloud-based software**: These solutions are less costly and easier to set up than building company servers and storage.

- **Visitor management system**: This digital system is used to register and track each guest, such as contractors, consultants, and vendors. Visitors can check in and check out via their mobile phones. Hosts are notified when visitors arrive and location-based services provide navigation.

- **Occupancy sensors**: Counting sensors or surveillance cameras can count people as they come and go through a defined area unobtrusively. Occupancy sensors placed on or near workspaces can provide real-time availability information.

- **Thermal screening**: Thermal cameras scan people quickly and can also provide accurate real-time occupancy counts.

While IoT sensors collect and measure data, a method of transferring that data to space management software will be required. The technologies to perform this transfer were outlined in detail in *Chapter 4, How to Make Buildings Smarter with Smart Location*. There are numerous off-the-shelf software solutions that can integrate space management with workplace management systems. Different building types and building purposes will help determine the software required. The following are some examples of space management in different building types:

- **Office environments**: IoT occupancy data provides actionable data about which rooms, offices, spaces, and floors are being used. From this data, office designs and layouts are determined, and HVAC adjustments can be made to control energy costs.

- **Retail**: Occupancy sensors can monitor how customers move around a store and which sections of the store they spend more time in. With this data, sales displays can be positioned, location-based coupons can be sent to a person's phone, and staff can be maneuvered to provide assistance.

- **Event space**: Using attendance records, heat mapping, and other occupancy collecting techniques, space can be reconfigured to match traffic patterns at conferences, sporting, and concert events.

- **Industrial**: Eliminate or reduce the need for plant expansion by optimizing a facility and its related processes. Using IoT sensors to collect valuable worker and production process data, pinch points can be identified and managed, an inventory can better be balanced, and space can be reengineered to be more effective.

- **Universities and school buildings**: Occupancy and motion sensors can provide data to assist in determining the best classroom or event room layout to accommodate social distancing requirements and proper classroom sizes.

These are just a few high-level examples of space management in different types of buildings. Others on the long list of buildings include hotels, stores/shops, restaurants, labs, healthcare facilities, campuses, apartments, factories, warehouses, and more. Different buildings present different types of challenges and IoT-based solutions.

Financial interactions

Commercial real estate, retail, office park, multifamily, residential, flex space, and other building types typically will have a building owner/operator and tenant/renter relationship that requires financial transactions to occur. In some cases, it is a simple, single, all-inclusive monthly agreed amount to be paid by a certain date. More often, there are several components that comprise the monthly amount to be paid, with some fixed items and some variable items tied to usage.

This list may include a base rent or lease amount, utilities, parking, point-of-sales amounts for in-building purchases, amenity services, homeowners' association fees, tenant improvements, fees for common area use and maintenance, storage and delivery fees, and large-scale building repairs or purchases.

Building operators implement comprehensive solutions to manage all aspects of their interactions with tenants, guests, and potential tenants. Vacancy management, rent payments, maintenance cycles, and community association features are integrated into property management solutions. Customizable tenant apps empower the occupant with information and controls.

Residential building operators can implement smart remote management solutions. For example, when a resident moves out, building management can change the status of the apartment from occupied to vacant. Once placed in vacant mode, access codes can be changed, energy savings mode turned on for HVAC and lighting, and work orders generated for turn requests. Smart door locks and doorbells can be reset and used as security cameras throughout the building. Vacancy listings can be synchronized across listing partners such as Trulia, Craigslist, Zillow, Apartments.com, and others.

Property management software allows building operators to track leases and work orders, collect rent, and manage operation finances. Smart locker systems and automated package storage systems streamline mailroom operations, improve security, and create an audit trail, while providing immediate notification to an occupant. Occupants can command robots to deliver a package upon request. Smart storage delivery is often cited as the number one amenity to enhance an occupant's experience.

Important building operations information is collected from IoT sensors, devices, cameras, and controllers, and the data is used to calculate and document actual outcomes and results. This data can be used to meet the requirements established by government subsidy programs. Tax reductions, credits, government subsidies, and other financial incentives are often tied to real results that must be well documented.

Smart metering for gas, electricity, and water ensures each tenant is billed according to their actual use. Tenants can implement conservation practices that will directly impact their bottom line. Building operators can simplify their activities by transferring the billing, payment, and processing for each tenant directly to the utility company. Smart metering can also detect problems with irregular use such as a water leak that, if it had gone unnoticed, might have caused other damage throughout a building.

Data analytics

Data Analytics (DA) is the process of analyzing data collected from IoT devices using specific analytic tools and techniques. The intent is to turn large amounts of unstructured data into valuable and actionable information to make sound business decisions. This includes identifying trends and patterns from historical and current data, which is then used to make predictions and adjustments. IoT DA is now being used in every industry, including smart buildings, and is considered a subset of big data analytics.

Data Analytics Types

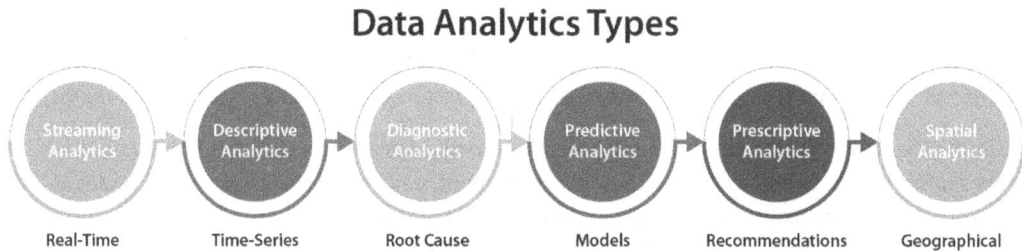

Figure 5.3 – DA types

Collecting data and performing IoT analytics can be broken down into the following types of analytics:

- **Streaming analytics**: Real-time data streams are analyzed to detect any urgent situations that may require immediate action. Also referred to as event stream processing, IoT building applications include water, gas, fluid leak detection, air quality monitoring, fire and smoke systems, and secure financial transactions.

- **Descriptive analytics**: Historical information is gathered and analyzed to review what took place in the past, when it took place, and how often it occurred. This is helpful to answer specific behavior questions and is used to detect any anomalies. Descriptive analytics can also be referred to as time-series analysis.

- **Diagnostic analytics**: This takes descriptive data to another step to answer the question of why a particular event or action occurred by identifying the root cause. Data mining and statistical techniques are used to identify hidden patterns that offer information about the cause of a problem.

- **Predictive analytics**: Using historical data and trends, IoT analytics predicts future events. **Machine Learning** (**ML**) and statistical algorithms are used to build models that are used to make predictions. Inventory management, maintenance, and demand forecasting are typical uses of this type of analytics.

- **Prescriptive analytics**: Taking predictive analytics to an advanced level, prescriptive analytics not only predicts what will happen but also makes recommendations on what should be done to correct it. This analytics uses optimization algorithms to determine the best action(s) to be taken to achieve specific goals.

- **Spatial analytics**: By analyzing geographic patterns, the spatial relationship between physical objects can be determined. These are used for location-based IoT applications such as asset tracking and smart parking.

Building owners and operators embrace DA to optimize operational efficiency, reduce costs, improve safety, and enhance an occupant's experience. Other benefits include real-time data analysis, improved scalability, increased accuracy, and improved security.

To build an effective data analytics program, we recommend starting by building a use case to identify the specific needs of an organization, determining an appropriate approach, and selecting the best analytics platform. IoT sensors and devices will need to be installed to collect data and data-cleaning processes will need to be implemented to ensure it is accurate and reliable.

Once data has been collected, it will need to be stored in a central database, which could either be locally on servers, computers, and edge devices, or a cloud-based storage platform. To understand and interpret the data, data visualization tools are required, and the data will need to be structured or semi-structured. Data analysis is the desired outcome and is completed by using different types of tools and techniques. Data analysis methods include ML, statistical analysis, predictive analytics, and **Artificial Intelligence (AI)**.

ML identifies and learns trends, patterns, and relationships between data points. Using past instances and experiences, it can adapt to change. **Natural Language Generation (NLG)** takes the ML findings and converts the data for automation and visualization.

AI techniques were traditionally applied to historical data analysis. The demand for real-time data analysis is driving scalable AI to enhance learning algorithms, making data more interpretable and more quickly. Augmented analytics is the combination of AI and analytics used to find hidden patterns and trends in large datasets.

The following are just a few of the ways DA is being implemented in buildings to reduce maintenance time, improve energy efficiency, and improve the occupant experience:

- Identifying broken equipment.
- Identifying air damper(s) that are permanently outside
- Identifying wasted periods of heating and cooling simultaneously
- Comparing energy use across a building portfolio and determining any outliers
- Identifying any sensors that may require calibration
- Analyzing energy use and predicting future requirements for budgeting
- Comparing similar equipment for benchmarking, performance, and trends analysis
- Comparing current results to *typical* patterns and results

Workforce applications

IoT technology can help improve operational efficiency and employee productivity. We have covered how IoT assists a workforce in building access control and security, along with how these devices provide valuable real-time data. Workforce management solutions combined with IoT-connected networks provide access to critical company data to ensure the workforce is distributed efficiently and effectively. Smartphones, wearables, and sensors collect data used by managers to allocate resources and track staff performance on a task-by-task basis.

Workforce Management (WFM) is software that streamlines and automates processes related to organizational performance levels and competencies. It is designed to optimize a worker's productivity and improve efficiency. These systems will track work schedules, time and labor, and workforce absences, ensuring payroll accuracy. They also organize environmental health and safety information collection, reporting, and root cause analysis. These systems reduce manual processes to improve workforce productivity and improve safety. Task management provides insight into employee skills for optimal task assignments.

Popular applications in this field include the following:

- **Attendance and timesheets**: IoT beacon sensors and smart tags automatically mark your attendance upon entering and exiting a building. Face recognition and biometric systems are touchless technologies on the rise post-pandemic for a safer workplace. These systems can also be used to monitor student class attendance.

- **Meeting and training registration**: Attendance and participation in key required meetings (that is, safety meetings) and training sessions can be immediately registered, and the individual's training records automatically updated to reflect the actual hours spent.

- **Health monitoring**: Wearables, smartphones, heat sensing, and sensors that can screen heartbeats, temperature, and other basic parameters can help identify and stop the spread of disease.

- **Employee tracking and monitoring**: You can monitor time spent and movement to reduce unproductive hours. When used in a hospital environment, the workforce can be precisely located in multi-story buildings, along with patient bed movement tracking and real-time location.

- **Employee empowerment**: Connected devices and networked systems provide employees with critical real-time information required to perform their job. Digital twin representation and the digitalization of user documents facilitate access to the real-time information needed to perform tasks. While this is beneficial in any building, this can be a lifesaving difference in hospitals.

- **Employee recognition**: Data collected and analyzed can be used to accurately monitor and measure **key performance indicators (KPIs)**. Employees and departments can be recognized accurately and timely.

Effective WFM systems need to be able to collect workforce statistics regarding worker performance in real time. They require the capabilities to analyze the current situation to provide improvement recommendations. A technology-driven WFM system integrates IoT to increase operations efficiency.

Concierge services and the amenity marketplace

Office amenities have a significant impact on tenant satisfaction, and from my experience, almost three-quarters of tenants wish their office building had more or better amenities. Fast Wi-Fi and free coffee were once thought to be an amenity but are now considered standard requirements. Smart buildings use IoT sensors and smart building applications to build a tenant portal with access to amenity services. These services help to attract and retain high-quality tenants and provide a safe environment.

The following are some of the many amenity services that benefit from using IoT smart building solutions to enhance the quality of a tenant's experience:

- **Conference facilities**: While most tenants will have meeting rooms in their space, they may not have space large enough to accommodate large events. Adjustable conference rooms with online scheduling and payments, along with smart tenant controls for temperature, lighting, and services, are also required. Smart conference rooms would include the items listed in *Figure 5.4*.

- Online Reservation
- Automatic HVAC and Lighting
- Smart Glass and Privacy Glass
- Smart Windows with IP
- Collaborative Conferencing Technology
- Audio and Visual Presets
- Presentation-Ready
- Conference call Autocall
- Shared Data
- Airplay
- Smart Jamboard
- Wireless Broadband
- Auto-Transcription
- AI Tools

Figure 5.4 – Smart conference room – smart amenities

- **Parking**: Finding parking spots is a real hassle when they are not assigned. IoT sensors can provide real-time availability to a driver to locate a spot quickly and safely. There are three types of sensors used – ground sensors, overhead or camera-based sensor technology, and simple counter systems that count the number of vehicles and open and close the gate based on availability.

- **Electric Vehicle (EV) charging**: As more and more people acquire electric vehicles, EV charging space will become a bigger challenge. Many buildings simply do not have the energy capacity to add more charging stations; therefore, availability and scheduling will become critical. Knowing what spaces are available and being notified when your vehicle is fully charged so that you can move it can be simplified by using IoT sensors.

- **Adjustable ergonomic workstations**: IoT solutions can transform ergonomics. Smartwatches can provide an adjustable standing desk of a person's height and adjust it when the person approaches. Chairs with IoT sensors can adjust and notify the person when it's time to take a movement break. Other IoT wearables can notify a person to correct an injury-prone posture.

- **Lockers**: Smart lockers reduce the risks of theft, misplaced, or stolen packages or other assets. They reduce mailroom overcrowding to maintain occupancy sensing and safe distancing requirements. Office buildings, education campuses, and large complex centers can streamline their mail processes. Apartments and condo buildings add smart package rooms:

  - **Mail/package storage**: Smart lockers offer contactless delivery, safekeeping, and delivery and pickup for staff and tenants. Tracking and notification solutions are integrated with the lockers. 24/7 access allows for convenient safe pickup. Pre-configured or customizable lockers are available to accommodate package sizes and quantities in real time.

  - **Bike storage**: Bike parking rooms offer secure innovative bike parking to promote cycling and free up vehicle parking spaces. Secure access, cameras, and IoT sensors control who has access and real-time space availability.

- **Fitness centers, wellness spas, and yoga studios**: Typically, these small facilities have limited space and even more limited equipment. Occupancy IoT sensors can provide real-time availability and scheduling, while also notifying a cleaning crew when a person has finished. Indoor air quality monitors can provide real-time air quality scores and adjust air ventilation as needed.

- **Tenant lounge**: Employees need a space to enjoy breaks with activities and comfortable seating. Space availability, open ping-pong and foosball table notifications, and people counting could be provided in real time with IoT sensors and tenant apps.

- **Food, beverage, and dining options**: On-site cafeterias and restaurants allow occupants to remain in a building and help attract highly sought-after tenants. Occupancy and space availability information can eliminate queues, and app ordering makes it easier to order and deliver on-site.

- **Entertainment**: Voice-activated devices included in tenant spaces provide entertainment and interaction opportunities. Building and community event and calendar information can be displayed on video devices or voice-delivered to smart speakers in the form of daily updates.

- **Public transportation/cars/taxis/ride-sharing**: Real-time information can be shared between a transportation or ride-sharing app and a building's tenant app to create a seamless experience. Information about building location points for pickup and drop-off can be delivered through navigation and way-finding apps. Buildings can provide car-sharing platforms and designated parking spaces. Access to strategically placed exterior cameras can provide real-time arrival information for ride-sharing and taxis, providing increased security.

- **Retail**: Having some retail on site is beneficial to busy professionals. Banks, shipping/mailing centers, dry cleaning, and dry goods stores make it easier for occupants to complete errands. IoT occupancy sensors can provide information on real-time crowds to prevent wasting time queuing. Other smart applications can feature online ordering, a robot delivery to an office, and product availability information.

- **Concierge services**: Dog walking, errand running, gift wrapping, VIP arrangements, coffee pickup, and a long list of other concierge services are now considered essential to certain class-A buildings and residents. These services have demonstrated benefits and improved retention. Real-time location and information can be provided using IoT location technologies.

- **Hair salon, manicure, and massage services**: Similar to fitness centers mentioned earlier, limited space and limited equipment create scheduling and use challenges. Occupancy IoT sensors can provide real-time availability and scheduling, while also notifying a cleaning crew when a person has finished. Indoor air quality monitors can provide real-time air quality scores and adjust air ventilation as needed.

More and more creative amenity services will undoubtedly be developed to help differentiate buildings, and IoT smart technology solutions will continue to be invented to meet these services. Availability, occupancy, location, health and safety, and secure real-time information requirements will be the key drivers for development and adoption.

Customized applications

Smart building technology simplifies the capability of building owners and operators to develop and serve customized tenant applications on a building's tenant experience app. Since the app is already deployed, smaller companies within a building might decide to take advantage of the access by developing company-specific information and programs. In some buildings, we have seen tenants' employee safety manuals and human resource policy guides digitized and made available via a tenant application.

QR codes located around an office allow workers to scan an item, area, or equipment and receive user manuals, safety manuals, and work process guides, for example. Companies leasing space in a building should work with the building operator to create and deliver a company-specific survey on the building's tenant applications.

Gamification

The Merriam-Webster dictionary defines **gamification** as the process of adding games or game-like elements to something (such as a task) to encourage participation. If an employer wanted to change employee behavior, they could implement a gamification process whereby points are rewarded based on the desired behavior.

IoT-enabled Gamification (**IeG**) is the convergence of gamification and IoT. Using sensors, beacons, and tags on an employee's ID badge, the employer could reward points for recycling, fitness activities, use of public transportation, or even participation in lunch-and-learn sessions. These points could add up to reward prizes or cash at certain levels, or just provide bragging rights within a company. Gamification and IoT can work together to bring synergistic benefits with an endless list of smart service possibilities.

Gamification is becoming popular for energy conservation in buildings and to help educate people to become more environmentally friendly. IoT can track your energy-related activities and provide rewards for taking actions to use less energy, such as turning off lights and better utilization of temperature. Points could be removed for bad activities, such as leaving a door open or leaving lights on in an empty room. Teams and communities could be established to compete with other, and social media callouts would be encouraged to get others to challenge each other.

Another very popular application of IoT gamification is the encouragement of improved health and wellness. Smartwatch devices that measure vital signs can be synchronized with sensors in a building's fitness center to provide daily summary reports, progress bars, badges, scoreboards, and league standings. Free incentives can be provided if the user hasn't been active for some time.

Gamification can also improve customer engagement across different industries. Social engagement allows users to share their personal achievements with others to create awareness and take part in challenges. Customer loyalty programs can drive customer engagement by offering discounts based on the number of visits or sessions attended. Coupons, discounts, product information, and additional purchase recommendations can be delivered in real time as gamification rewards.

Custom experiences

Carnival Corporation has developed what they call **experience Internet of Things** (**xIoT™**), which focuses on enhanced guest-crew interactions to perfect the guest experience. Cruise ships are a perfect application for an IoT smart building type ecosystem, since they are contained environments that create their own electricity, telecommunications, and operations systems to create essentially a building or even a small city of its own.

Carnival's MadallionClass ships provide a free wearable to all passengers to participate in their connected experience. A network of sensors and edge computing devices are positioned throughout the ship to make the guest experience customized, simplified, and more personal. Based on the guest's location, crew members can align certain products, services, and entertainment experiences based on the guest's preferences.

Creative opportunities will continue to evolve, driving the demand for customizable tenant-specific applications. Asset management, inventory, worker tracking, and customized video calling are just a few of the opportunities. Tenant experience is a foundational topic for software and SaaS providers. With multi-tenant environments, developers will need to prevent one tenant from accessing another tenant's resources.

Summary

Many new tenant experience applications are being introduced that are capable of handling, automating, or even eliminating certain building-related work order requests. Those requests that are processed can be opened, tracked, and updated by the use of IoT sensors and devices throughout a building. The aftermath of the pandemic is seeing a significant rise in smart space management, using data collected from IoT sensors and devices, and the use of DA. Building operators can gain efficiency and lower costs using workforce applications. Building amenities, tenant services, and customized tenant apps can create new revenue sources while increasing tenant satisfaction.

Part 1 of this book has explored the numerous IoT smart applications that are possible, and without a doubt, that list will continue to grow. In the next chapter, we will kick off *Part 2* of the book by examining smart building architecture and the smart building ecosystem.

Part 2:
Smart Building Architecture

In this part, we review the technologies used to build a smart building from the edge routers, numerous **Internet of Things (IoT)** devices, the various connection options available, and the software required both locally and via the cloud to make it all work together seamlessly. Use cases and digital twin technology will highlight an architecture that brings the components together.

This part contains the following chapters:

6

The Smart Building Ecosystem

While many of the new buildings under design and construction today have already incorporated IoT and smart building solutions, smart buildings do not need to begin from the ground up. Existing buildings known as *the built environment* may add IoT technology and smart building applications to make a building smarter.

What are the components necessary to build an IoT network to create a smart building? We will review the major components of the smart building ecosystem: IoT sensors and devices, edge, fog, or cloud computing, data management, analytics software, a user interface, and a means of connectivity.

In this chapter, we're going to do the following:

- Explore the various IoT sensors and devices that monitor an environment and send data to a computer for processing and viewing

- Discover the differences between on-site computing and cloud computing and how smart buildings combine them

- Examine how information is transmitted from an IoT device or sensor to a computing platform, using wired or wireless IoT communication technologies and protocols

- Delve into the numerous smart building software and APIs that are available to process and analyze the data collected by sensors

- Learn how to manage, store, and process data from thousands of sensors

- Show how data is viewed after processing with a user interface of some sort

- Demonstrate the capability to send commands back to a device for action

IoT sensors and devices

Smart building devices are the systems and platforms that perform a specific task. These devices typically will have built-in intelligence to gather information and react accordingly. Many may have built-in connectivity as well such as cloud connectivity, BACnet, Modbus, or similar connectivity protocols. Examples include a building's HVAC system, smart elevators, perimeter access, door locks, cameras, readers, actuators, controllers, and many more. For systems and devices that do not have built-in intelligence and connectivity, IoT sensors or tags can be affixed and connected directly to them.

A smart building's IoT network begins with smart building sensors that collect operational and environmental data and automate certain activities. Monitoring and measuring a connected device are the inherent functions, along with noting any changes in the measured parameter and alerting when it's out of range. There are many IoT sensor types, and here we will focus on the most common sensors driving smart buildings today.

First, we have **environmental sensors**, which include the following:

- **Temperature sensors**: Smart temperature sensors measure an occupied space and determine what settings the HVAC system should be set at while providing real-time access and information. This single sensor can deliver the biggest contribution to smart building benefits by managing energy costs and improving occupant satisfaction. Temperature sensors are used to control computers and equipment from overheating, and to monitor items such as food and pharmaceutical storage that require very specific temperature ranges.

- **Humidity sensors**: Often, a humidity sensor is integrated or co-located with the temperature sensor in the same unit. The humidity sensor monitors and controls the HVAC system to manage safe levels to reduce mold, mildew, and viruses. Not only is occupant safety and comfort a focus for humidity, but most equipment and machinery are subject to specific tolerance levels to prevent rust and mold.

- **Air/water/gas quality sensors**: IoT sensors monitor and collect air, water, and gas environmental information, and computer technology calculates acceptable ranges for particulate matter, carbon dioxide, TVOC, formaldehyde, and many other factors. They can detect harmful, toxic, flammable, or combustible gases and shut down equipment while alerting building operations personnel.

- **Chemical sensors**: These sensors can detect chemical substances that may have accidentally leaked from their containers. This is extremely important in spaces occupied by people.

- **Water leak sensors**: Water leak sensors are valuable for real-time monitoring to prompt immediate action to reduce damages. Water leaks can occur not only from the building's plumbing system but also from the air-conditioning, HVAC, and manufacturing equipment that require water or measure water levels near critical equipment. These simple sensors can sound an alarm the second water is detected.

- **Smoke sensors**: Smoke detectors monitor and measure the concentration of smoke to prevent fires or notify personnel and other systems when there is a fire. Many methods for smoke detection are available, and one type uses infrared beams that are sent to a photosensitive tube, and if that beam is not properly measured or is blocked, smoke alarms are triggered. An early warning system with smart IoT smoke sensors can reduce damage and save lives by automatically managing airflow and equipment responses.

- **Electrical current monitoring sensor**: Electric flow sensors look to measure the continuous energy flow at the circuit, machine, or zone level. This allows a building operator to determine energy use and waste to decide on changes or investments that might be required.

- **Light sensors**: These optical sensors measure the amount of light in a particular area and usually have motion sensors integrated that, when activated, trigger a light to turn on, off, or dim.

- **Contact sensors**: These are also known as position sensors and are often used to detect an open or closed window or door using magnets. These magnets touch each other when closed to show a closed state and show an open state when the sensors are not touching.

Motion sensors include the following:

- **Occupancy sensors**: These sensors are used to determine how many people are in a particular space or building, and to determine in real time which desks are free.

- **Infrared detection sensors**: These sensors determine a person's presence and react by turning on lights or setting off an alarm. Other building applications include automatic doors, toilet flushers, and hand dryers.

- **Cameras and microphones**: These devices can provide a more specific and immediate sense of a situation or movement. Voice-command systems such as Alexa, Siri, and Google Home can also detect sounds such as a window breaking or active gunshots.

- **Thermal imaging**: These imaging cameras can check the temperature of individuals entering a building to detect motion.

Other smart building sensors include the following:

- **Proximity sensors**: These sensitive sensors can recognize objects and react accordingly. They are used for automatic doors, and a building's hot or cold air blowers. They are also used for anti-theft purposes, with tags affixed to assets and proximity readers strategically placed to set off alarms.

- **Pressure sensors**: Pressure sensors are used in safety control systems for compressed air, gas, or liquid. They are also used for leak testing.

- **Power-off sensors**: These are often used to measure a power line supply. Alarms are triggered, and alerts are sent when there is a power failure or disruption.

- **Accelerometer sensors**: These sensors are used for remote sensing applications such as a building shaking during an earthquake. They are used to monitor buildings, industrial plants, and structural monitoring.

- **Gyroscope sensors**: This is a control system that measures movement and gesture positioning within a space based on the speed of rotation. These are used to measure a building's vibration and for in-building navigation systems.

- **Level sensors**: Used in industrial, manufacturing, water treatment, beverage, and food manufacturing applications to detect the level of substances and liquids.

- **Image sensors**: These sensors take optical images used to display or store data electronically. Medical imaging, digital cameras, and biometric devices are just a few of the applications.

- **Health sensors**: These are infrared sensors that can read a person's temperature and blood pressure from a distance.

These and numerous other sensor types play a critical role in an IoT network by collecting valuable data and, in some cases, triggering immediate life-saving actions. Most smart building sensors are designed for both information collection and automation.

Connectivity

Collecting data with IoT sensors and devices would be pointless if that data could not be transmitted to the computing platform and applications for analytics and viewing. Connecting these points should be conducted without making changes to the applications or systems.

The application programs should be able to communicate with each other to complete the transaction. This communication system exchanges signals between destinations through a channel, and the process is referred to as network communication. A transmitter to send the information is required, along with a receiver at the other end.

The two types of communication systems are line communication systems and radio communication systems. Line systems use existing infrastructure or dedicated physical mediums such as wire, fiber, cable, coax, or similar to transfer data. Radio systems use radio waves for this transfer.

Communication systems are also classified as the following:

- **Analog communication systems**: Audio, video, and pictures are transferred using analog signals. Analog signals are electrical impulses or electromagnetic waves that use continuous time-varying signaling to transmit data.

- **Digital communication systems**: This involves the electronic transmission of encoded data such as internet communications.

In this section, we are going to cover the different methods used to transmit data, followed by the different industry-driven communications protocols used in all types and categories of buildings.

Transmission methods

For many decades, basic copper wiring was used for communication between building systems, but technology gains have opened and expanded possible methods. Often, the driving force for these new methods is the amount of data being transmitted and the speed required to achieve real-time requirements:

- **Wire**: Wired communication lines have been around since the first telegraph wire was introduced in 1844. The insulated telephone wire was introduced in 1876 but only for low-voltage applications. First, single grounded wires were used, but they were too noisy. Copper wires were used to reduce the noise, but the wire was not strong. Two-wire circuits eliminated much of the electrical disturbance and noise. Paper was added to wrap each wire to further eliminate noise, and post-WWII, polymeric materials replaced the paper. Single copper wires evolved into twisted pairs, and sheathing was added later on. Today, there is a large variety of communication wire types, sizes, and bundles available for any communications requirement, and they are classified according to design, transmission range, and frequency.

- **Cable**: A data communication cable is an insulated conductor wire used to carry data, video, and voice between devices. Without digging too deeply here into the various types of cable, we will summarize that there are unshielded, shielded, twisted-pair, and coaxial cables that fit into this category.

- **Ethernet**: Ethernet cables are like a traditional phone line, but they are larger and have more wires. Ethernet cables contain eight wires, while phone lines contain four wires. This popular inbuilding cable is used to connect devices, computers, switches, routers, and more, but it is limited by its length and durability. These cables plug into Ethernet ports on a device. The most common industry standard ethernet cables are known as **Category 5 (CAT5)** and **Category 6 (CAT6)** cables. These cables can contain stranded or solid wire, with the latter offering better performance and stranded cables being less susceptible to physical cracks and breaks.

- **Power over Ethernet (PoE)**: This uses the Ethernet cables described earlier, with the difference being that this method allows communications data and electrical power to be transmitted over the same cable at the same time. This allows a single cable to be run to an IoT device to supply power and carry out network communications.

- **Fiber**: A fiber optic cable contains strands of glass fiber wrapped in an insulated cable. Fiber is best applied for long-distance and high-performance applications and is less susceptible to interference, but it does cost quite a bit more than wires. It provides higher bandwidth and can transmit further than wired cables. Today's desired in-building fiber application consists of a horizontal and vertical distribution method to scale each floor in a building and to reach across each floor.

- **Wireless**: Data is transmitted in the form of electromagnetic waves through free space and connects with devices that have permission to connect. No physical conductors are used for transmission. These wave signals can travel through the lowest layers of the Earth's surface using low frequencies (ground propagation) or by using high-frequency radio waves transmitted over larger geographical areas (sky propagation). A third method known as line-of-sight uses very high frequencies to travel in a straight line from the source to the receiving device. These waves can be disrupted by obstacles in their path.

Later in this chapter, we will review the most common wireless protocols used for in-building applications.

Traditional building communications protocols

A communication protocol is a set of rules that define how two or more entities transmit any variation of information within a communication system. Syntax, semantics, synchronization, and rules of communication are defined in a protocol, along with possible error recovery methods. These rules apply to hardware and software that communicate with endpoints or devices.

	BACnet	LonWorks	DALI	KNX	Enocean	Zigbee	MQTT	AMQP
Applications	HVAC, lighting, security, and fire systems	HVAC, lighting, process control, and automation	Lighting, motion detectors, and gateways to other protocols	HVAC, lighting, remote access, security, and energy management	Occupancy sensors, key cards, lighting controls, and other room controls	HVAC controllers, room controllers, and occupancy sensors	IoT messaging, HVAC controllers, occupancy sensors, and other room controllers	IoT messaging, HVAC controllers, occupancy sensors, and other room controllers
Developed and supported by	ASHRAE	Echelon Corporation	Philips	Konnex Association	Siemens AG	Zigbee Alliance	OASIS and ISO Standard	OASIS and ISO Standard
Type	Wired and Wireless	Wired and Wireless	Wired and Wireless	Wired and Wireless	Wireless	Wireless	Wired	Wired
Medium	Twisted-pair, wireless mesh, and fiber optic	Twisted-pair, wireless mesh, fiber optic, and power lines	Single cable pairs create the network bus	Twisted-pair, radio frequency, IP/Ethernet, and power lines	Wireless	Wireless	Hardwired	Hardwired
Transmission Mode	IP, Ethernet, LonTalk, ARCnet, Zigbee, and MS/TP	Predictive p-persistent CSMA	Gateways	Gateways	Carrier Sense Multiple Access (CSMA) with collision detection	TCP or UDP	TCP/IP	TCP/IP
Security	Transport layer security (TLS) and Open Authorization (OAuth)	No data encryption. Implements sender authentication	No security measures	Implements data encryption and authentication	Encrypted data using AES algorithm with 128-bit key	Encrypted data using AES algorithm with 128-bit key	TLS encrypted messaging and authentication, OAuth	Integration of TLS and Simple Authentication and Security Layer (SASL)

Figure 6.1 – A comparison of traditional building communications protocols

Figure 6.1 provides a comparison of the traditional smart building communications protocols, and each one is described as follows:

- **BACnet: Building Automation and Control Network** (**BACnet**) is by far the most popular communication protocol used by buildings around the world. It was developed and supported by the **American Society of Heating, Refrigeration, and Air-Conditioning Engineers** (**ASHRAE**). BACnet has defined five interoperability levels: device and network management, trending, data sharing, scheduling, and alarm and event management. It is used for both wired and wireless applications. When endpoints are connected to the BACnet network, users and systems have direct access to manage and control that endpoint or device.

- **LonWorks: Local Operating Network** (**LON**) is also referred to as **LonWorks**. It is a proprietary protocol that consists of neuron chips from various vendors, the LonTalk protocol, physical connectors (wires and cables), and network management tools. Developed for building automation, it is very popular and is used around the world. LonMark International supports and promotes the protocol and its related standards. It is used for both wired and wireless applications.

> **Note**
> While BACnet and LonWorks have similar overlapping scopes, they are not interoperable.

- **Modbus:** While not often used specifically within a building's infrastructure itself, this communication protocol is used mainly for connecting industrial electronic devices. Developed specifically for industrial applications, it uses serial communication lines or Ethernet protocols. We have not included Modbus in the side-by-side comparison in *Figure 6.1*, since it has limited application within buildings.

- **Digital Addressable Lighting Interface** (**DALI**): This was developed for building automation systems and lighting controls. This communication protocol is bidirectional so that a real-time operating state can be obtained. 1 to 64 devices can be connected per DALI network, and multiple DALI networks can be connected. This interface is used for both wired and wireless applications. Lighting supports 256 levels of brightness.

- **KNX**: BatiBUS, **European Home Systems Protocol** (**EHS**), and **European Installation Bus** (**EIB**) standards were combined to create the KNX international building automation standard. It supports point-to-point and multicast communication. With three levels, easy, system, and automatic, it works on tree topology whereby each node is related in a hierarchy, and KNK is supported by the KNX Association. It is used for both wired and wireless applications.

- **EnOcean:** This wireless communication protocol was developed specifically for buildings and energy-harvesting devices that do not require a power source or batteries. Nearly 1,000 products have been developed and certified for use with this protocol. This low-cost and flexible protocol uses thermal and kinetic energy harvesting techniques for a greener and more cost-efficient solution.

- **MQ Telemetry Transport (MQTT)**: This was designed as a machine-to-machine communication protocol for remote locations with resource-constrained devices or poor network bandwidth. This bidirectional publish/subscribe messaging transport is becoming the standard for IoT messaging. MQTT is best suited for high-latency, low-bandwidth networks with large numbers of embedded devices. MQTT-enabled devices/sensors can directly connect with the cloud.

- **Advanced Message Queuing Protocol (AMQP)**: This is an open standard communications protocol designed for message-oriented middleware. Another very popular IoT messaging protocol, AMPQ is best situated for large enterprise-wide systems that require highly customized and involved messaging with specific reliability, security, and interoperability requirements.

Open communication standards are evolving to provide better connectivity and interoperability among connected devices replacing proprietary protocols. Open standards allow third-party devices to connect to a network easier. With the number of IoT devices growing significantly in buildings, non-traditional communication protocols are emerging, mostly in the form of wireless applications.

Wired	Wireless
BACnet 🖧 BACnet 1987 – this low cost, has no licensing fees, and is used to communicate between building devices. Defines 60 standard object types. The protocol services include Who-Is, Who-Has, I-Have.	**Wi-Fi** 📶 Wi Fi With wireless internet available, Wi-Fi is one cost-effective and easily accessible way to connect IoT devices. Drawbacks include interference, limited bandwidth due to many connected devices, and the amount of power it requires. Used for thermostats, lighting, smart devices, and broadband internet access.
Modbus 🕸 Modbus 1979 – as a communications protocol, this is a common means of connecting electronic devices. It is low-cost, has no usage fees, and is used in HVAC, lighting, life safety, access controls, transportation and maintenance.	**Bluetooth** 🅱 Bluetooth 1989 – Bluetooth uses radio waves to communicate. Bluetooth devices contain computer chips with radios to allow everything to talk to each other. Hundreds of products are compatible with Bluetooth automation. The main drawback is its range restriction.
LonWorks 📶 LonWorks 1990 – designed as a low-bandwidth protocol that supports five communications media: twisted pair, power line, radio frequency, coaxial cabling, and fiber optics. It is the highest costs and comes with licensing fees.	**Zigbee** ✔ ZigBee 1998 – a protocol created specifically for commercial use, this is the most widely used for building automation. It uses a mesh network to create long ranges and fast communications via radio frequency with minimal power usage, lasting several years on a single set of batteries.
MQTT 📶 MQTT A lightweight, publish-subscribe network protocol that transports messages between devices and usually runs over TCP/IP. Designed for connections with remote locations or where the network bandwidth is limited.	**LoRa** LoRa A proprietary low-power, wide-area network modulation technique. It is based on spread-spectrum modulation techniques derived from **chirp spread spectrum (CSS)** technology. **Near-field communication (NFC)** 📶 A set of communication protocols that enables communication between two electronic devices over a distance of 4 cm (1 $\frac{1}{2}$ in) or less. NFC offers a low-speed connection through a simple setup that can be used to bootstrap more-capable wireless connections. **Narrowband Internet of things (NB-IoT)** 🌐 NB-IoT A low-power wide-area network (LPWAN) radio technology for cellular devices and services. Focuses specifically on indoor coverage, is low cost, and has a long battery life and a high connection density. NB-IoT uses a subset of the LTE standard.

Figure 6.2 – Common IoT communication protocols

BACnet, Modbus, and LonWorks have been the most widely used communication protocols, largely based on their wired backgrounds, their use with traditional building systems, and the amount of time they have been available. Recently, we have seen greater use of MQTT for IoT deployments, and while the traditional protocols have added wireless capabilities, other wireless protocols listed in the next section are typically easier to deploy and use and oftentimes less expensive, making them better suited for IoT networks.

Common wireless IoT building communications protocols

There are numerous wireless communication protocols used in buildings for IoT communications. At one time, we counted over 31 different wireless protocols, and that number continues to grow. Here, we will review the most commonly accepted protocols for smart buildings:

- **Wi-Fi: Wireless Fidelity (Wi-Fi)** is extremely popular for **wireless local area networks (WLANs)** and IoT networks utilizing the IEE 802.11 standard. Devices typically need to be within 20 to 40 meters of the source, with data rates up to 600 Mbps. Wi-Fi is easy to deploy and has low costs; however, power consumption is high, and the Wi-Fi range is very moderate. Wi-Fi 6 is the latest iteration and provides faster speeds, better security, and increased battery life over previous versions.

- **Bluetooth**: Another very popular protocol that is used over short distances is Bluetooth. It uses short-wavelength UHF radio waves on the 2.4 GHz band. Depending on the application, there are three different versions. Bluetooth is used for IoT and M2M devices. **Bluetooth Low Energy (BLE)**, also known as **Bluetooth 4.0**, uses less power and lasts longer than Bluetooth. iBeacon is used by Apple to communicate between iPhones. The Bluetooth range is 50 to 150 meters, with a maximum data rate of 1 Mbps.

- **Zigbee**: This low-power building automation protocol is based on the IEE 802.15.4 standard and is a mesh network topology. Because it is a mesh network, if one link is down or broken, another link can be automatically used. Its long reach is great for large, tall buildings, and campus environments. Each Zigbee device can support 240 applications and endpoints, with a range of 10 to 100 meters.

- **LoRaWAN (Long Range Wide Area Network)**: Fast becoming one of the most popular protocols, LoRa was designed as an IoT network protocol. It is just as its name implies – a low-power, long-range protocol capable of connecting millions of devices with a range of 2 to 5 km.

- **Near-Field Communication (NFC)**: This enables two-way interaction between devices such as contactless payments using a smartphone. For smart building applications, it is used to access digital content and connect electronic devices. It has a very limited range of about 10 cm.

- **Narrow Band Internet of Things (NB-IoT)**: This is LPWAN technology using low-power and long-distance communication with a large battery life. Its signals can transmit in underground areas and through walls where cellular signals cannot reach. NB-IoT can travel up to 10 km.

- **Cellular and 5G**: Cellular is used for long-distance communications and can send large amounts of data at high speeds. GSM/GPRS/EDGE(2G)/UMTS and HSPA(3G)/LTE(4G) are the cellular communication protocols that have been used for the last 2 decades. **Fifth-generation (5G)** cellular is the latest cellular release and is designed to transfer data at high speeds between IoT devices and smart devices.

- **Radio-Frequency Identification (RFID):** This uses electromagnetic fields to locate and identify tags and objects. These tags and devices contain stored information about a device. It has a range of 10 centimeters to 200 meters.

- **Z-Wave:** This low-power RF communications protocol was designed specifically for home automation products and uses a mesh network with low-energy radio waves to deliver a longer battery life. Smart devices such as smart door locks, garage door openers, lights, and thermostats are common applications of the Z-Wave communication protocol.

No single wireless communication protocol is the best, nor is there one that is right for every deployment. Unique circumstances and a range of factors such as power needs, device locations, geographic size, features to be deployed, security requirements, and other applications should all be considered when deciding what protocol to use.

Three layers of computing

We have established that IoT sensors collect data, and that data is transmitted using a communication protocol to a computing platform of some sort to be analyzed, viewed, and acted upon. There are different methods to achieve the computing process based on the amount of data, the speed and time required, and the financial resources available. This computing activity may take place locally within a building on or close to the source, in the cloud, or through a combination of the two.

Building management systems, building automation systems, security systems, fire alarm systems, and other building systems have traditionally been used to build computing platforms, albeit as separate and disconnected systems. Basically, each of these systems has a computing platform of its own, either within the device itself, on a dedicated server, or on a personal computer. More recently, buildings have added cloud computing in an attempt to connect these building systems to aggregate, analyze, and visualize holistic data.

Edge and fog computing

Edge computing facilitates data processing and analysis close to the source where data is collected and not on the computing servers or cloud. This provides for much faster response times, larger volumes of data, and less reliance on internet bandwidth and connections. In some cases, there can be cost efficiencies gained from local computing as well as greater levels of security.

In most cases, the computing occurs directly in or on the devices the sensors are connected to. The simplest example is the security and privacy features, such as encryption and biometric processing, located within smart devices and smartphones. Data that is processed locally means there is less data that must be transmitted to the cloud, thereby reducing costs while increasing response times.

Edge computing keeps data and processing on the sensing device itself to eliminate the need to send data to the cloud. On the other hand, **fog computing** aggregates and processes data collected from multiple sensors and transmits only the *important* data to the cloud.

Both edge and fog computing use local networks and computing capabilities to perform the tasks that otherwise would be performed in the cloud. Fog improves upon edge's capabilities, since it offers greater capacity to handle more data at once.

Using video security surveillance as an example, a small building's one-camera data could easily be streamed to the cloud because it requires little bandwidth. For a large building with multiple cameras, large amounts of data would be collected, creating transmission bottlenecks and high transmit costs. The data from multiple cameras can be aggregated and processed locally on a fog computing network and then only key footage and information are transmitted to the cloud.

Processing and computing data near its source with edge and fog computing provides several advantages:

- **Latency**: Measured in milliseconds, latency is the delay time that occurs on signals when transmitted between devices, computers, and the cloud. Critical building systems and processes cannot afford delays that occur when large amounts of data from sensors are sent to the cloud and an immediate response is required. Some examples included facial recognition sensors, leak detection, and smoke alarms.

- **Bandwidth**: If we think of internet connections as pipes, we can then recognize that there are limits to the amount of data that can be transmitted through these pipes at any given time. Bandwidth defines the amount of data that can pass through. Fog and edge computing reduce the amount needed as well as the cost to transmit data, as less bandwidth is required.

- **Load balancing**: Fog computing servers can route data between servers most efficiently in terms of capacity and speed.

- **Security**: Security vulnerabilities are reduced when data is managed on-site. Fog computing servers can monitor incoming traffic and block anything that might be dangerous and keep hackers out by providing secure firewalls. On-site conditions and requirements such as locking doors, setting off alarms, and closing valves can be performed in milliseconds.

- **Control**: Fog servers provide more controls and better decision-making functions.

Fog and edge computing both bring computing capabilities closer to the source of data, which allows information to be processed more quickly. Both are relatively new technologies that are opening the door for new and exciting use cases, due to their efficient data transfer, reduced latency and costs, and real-time computing capabilities.

Cloud computing

Cloud computing provides access to very powerful computing capabilities without expensive hardware. The computing is conducted on a network of remote servers that are hosted on the internet. From there, data is processed, stored, shared, and managed. Building operators have access to IT infrastructure, hardware, and software resources instead of buying and supporting their own local servers and personal computers.

Third-party companies host and manage remote high-end server networks and computers that make up the cloud. With the cloud model, users have access to a much more powerful and efficient computing platform than they could possibly have with their own technology. It increases computing capabilities without adding infrastructure and software costs.

The cloud system has a *frontend* that includes the user's interface, such as computers, networks of computers, smart devices, phones, and other similar devices. The *backend* contains the cloud storage systems (that is, servers), hardware, and software for data processing. The internet is used to connect the frontend and backend.

There are many scalable cloud services:

- **Infrastructure as a Service (IaaS)**: Buildings have access to services that would traditionally be located in the on-site computing center. These services can include data backup, data recovery, and load balancing.

- **Platform as a Service (PasS)**: Building programmers can access computing platforms, computer languages, and other tools to develop and code their own applications.

- **Software as a Service (SaaS)**: These are on-demand, turnkey, web-based applications such as email, conferencing, project tracking, and time management, among others.

While these computing technologies may differ in purpose and design, they often complement each other and are used together. Edge and fog computing offer decentralized data management for IoT needs and require real-time processing, accessibility, and security. Cloud computing allows the scalability of constantly growing IoT data that can be distributed quickly and accessed by connected devices anywhere.

Software and APIs

Many of the OEM vendor IoT devices such as building management systems, building energy systems, and building automation systems have computing capabilities contained within a system. Custom and often proprietary software computer programs compile and analyze data, provide visualization methods, and deliver commands to physical devices such as controllers and actuators to perform an action, such as opening or closing an air damper.

In a smart building environment, our goal is to connect and control many of these systems and to give building owners, operators, and occupants direct access and control within perimeters. Many of the applications we discussed in part one of this book require customized software to collect and analyze data and perform any specific actions required. Like smart home applications that control lighting and temperature, specific software programs are needed to operate smart features in a building.

Often, these application programs are single purpose, and hundreds of different software programs are required to manage an entire building. SaaS smart building solutions are popular among facility managers to improve operational efficiencies and productivity. Hundreds of SaaS companies are developing unique applications and now competing in this space. While the list is long, some of the more popular solutions are Cohesion, Particle Space, Comfy, HqO, Tapa, Apache Hive, Modo Labs, Enlighted, Building Engines, and Acuity.

Software programs have been developed as smart building platforms that allow individual smart solutions to come together in a holistic manner. The following are some of the leading solutions:

- **Microsoft Azure**: Azure is a multi-level platform that allows smart building solution partners to ingest building equipment and environmental sensor data, to integrate with their solutions using Azure IoT. Azure spatial intelligence capabilities can build models.

- **Intel IoT platform**: Smart building pre-validated building blocks, development tools, reference architectures, and design resources are made available to facilitate software development.

- **Google Digital Buildings**: Hosted on GitHub, this is an open source uniform data schema to represent buildings and building-installed equipment such as HVAC. It provides a simple configuration language and validation tooling.

- **AWS**: AWS IoT TwinMaker supports developers in the creation of a digital twin for the building. Amazon Alexa is integrated into many of the new IoT devices to listen and facilitate command controls. AWS has numerous application and solution partners hosted on its network.

- **Cisco**: Cisco's hybrid cloud can build and operate solutions for smart building operators with a single-managed IT network infrastructure. It has a suite of hardware and software solutions such as Cisco DNA Spaces and Digital Brew.

- **Tapa**: Tapa is an edge server and software that connects directly to a building's communication protocol (typically, BACnet) and gives building operators direct local and cloud access to IoT devices, sensors, actuators, and controllers to manage a building and related access to data.

The proliferation of smart building applications has been made simpler mainly through improved collaboration by using APIs. An **Application Programming Interface** (**API**) allows software development companies to open their application's data and functionality to other partners and developers, establishing communication with each other and building their own platforms with a customized interface.

This lets products and services communicate with each other through a documented interface to access each other's functionality and data. How an API is implemented remains confidential. A common example of an API is a website allowing an individual to log in using their Facebook, Twitter, or Google profile login information.

Data management

In *Chapter 5*, *Tenant Services and Smart Building Amenities*, we described data analytics as the process of analyzing data collected from IoT devices using specific analytic tools and techniques. We reviewed the different types of analytics and provided examples of how data analytics can be implemented in buildings.

Building data traditionally has been collected and stored in system silos, which makes it difficult to analyze and manage. The goal of a smart building is to provide a holistic view of the entire building's operation, and therefore, this data needs to be integrated to provide an exchange of information. To achieve this, massive amounts of data need to be ingested and processed, a task that is too large and complex for traditional database systems.

Big Building Data (**BB Data**) is a dedicated big data system for processing, sharing, and storing all a building's data. Input is collected from different sensors using a common method and language for compatibility. Batch processing is performed while the data sources are at rest, and real-time processing is performed while data is in motion. Predictive analytics, machine learning, and artificial intelligence are used to perform advanced analytics due to the massive amounts of data. It can extend from one building to another to provide a portfolio view.

Here are the key pillars of a big data platform:

- **Data management**: As the saying goes, what gets measured gets managed. Big data begins with the sourcing of data from relational databases, files produced by the applications, and real-time information from IoT sensors. Data ingesting is the process of acquiring data from different sources. Data integration is the process of integrating disparate data into data lakes or repositories. Data engineering transforms data to be used downstream.

- **Data modernization**: This process transfers siloed data from legacy databases to more powerful and advanced data lakes or cloud-based data storage. Data lakes store unstructured and structured data and can hold large volumes of data in different formats for batch processing.

- **Data science**: Since datasets are very large, big data solutions process batch jobs to filter, aggregate, and prepare data for analysis. Cognitive automation uses applied artificial intelligence and machine learning solutions to provide descriptive insights, predictive analytics, and actionable insights.

- **Data consumption**: Data is prepared and served in a structured format for use by analytical tools. Data is displayed in graphic representation for data visualization to meet the goal of providing insights through analysis and reporting. Dashboarding allows building operators to monitor performance in real time. Report migrations consolidate reports to provide valuable insights.

- **Data orchestration**: Data solutions typically consist of repeated processing operations and workflows. Orchestration technology allows workflow automation and pushes results directly to a dashboard or report.

- **Data cleansing**: Sometimes referred to as data cleaning, this involves identifying, locating, deleting, or correcting inaccurate, corrupt, irrelevant, or incomplete data from databases, tables, and record sets.

There are many challenges in attempting to implement big data. Many of the components are distributed on many different machines. The learning curve is steep, and maintenance is intensive. Technology continues to evolve rapidly, with new solutions introduced constantly and the number of connected devices growing daily.

More often, the term *big data* refers to the value extracted from the data and through analytics, and not the size of the data. Smart cities require reliable data as the basis for making decisions, and much of the sensor data can be acquired from smart building big data.

User interface and single pane of glass (SPOG)

IoT devices collect data, and a communications network transfers that data to a computing platform locally or on the cloud. At this point, individuals will need to interact with the data and the computing platform to monitor and control building systems. The **User Interface** (**UI**) is the defined method of interaction between an individual and a computer. This interaction is conducted in the form of screens, pages, forms, buttons, and other visual elements.

Simplicity and efficiency are the desired goals for this interaction, and users typically prefer to use intuitive UIs. These replace the need for the user to learn specific commands and languages, thereby allowing non-technical people access to data. Often, users are alienated by poorly designed or overly complex navigation applications. Good UI typically hides the complexity behind the interface.

Types of UI include the following:

- **Graphical User Interface** (**GUI**): Audio indicators and graphical icons (pictures, icons, and menus) are used to interact with electronic devices instead of text.

- **Command-Line Interface** (**CLI**): Text commands are used to execute operating systems. Developed in the 1960s, computer terminals were used to interact with computers. Later in the 1970s and 1980s, Unix and PC systems used CLI, which today is used to install software and access other non-graphical features.

- **Menu-driven interface**: Like its name, this system lists menu choices that can be selected to navigate an application. Kiosks and ATMs are examples of menu-driven interfaces. Little to no training is required and less computer processing power is required.

- **Form-based interface**: Text is entered into fields by choosing one of several values. For example, word processing software offers text lists to select the font type and text size. An example is an online job application form with drop-down menus for data selection.

- **Natural-Language User Interface** (**LUI** or **NLUI**): Linguistic verbs, clauses, and phrases act as the control for selecting and changing data. A key goal is to create an interface that is an intuitive natural language.

Often, the design of a UI can determine whether it's successfully adopted by a user. When developing the UI, developers should keep the interface simple by using common elements that users feel comfortable with. Attention should be drawn to the placement of items and the use of the right colors and contrast. Analytical applications and platforms that use different charts and graphs help the user to understand consumption trends, anomalies, and alerts. Anticipate the user's needs by developing pre-chosen fields and presenting the most searched items so that the user does not have to search.

SPOG

As smart building software programs and applications begin to accumulate, it becomes more difficult to determine how these different systems and data are related to each other. Multiple IoT data sources, each having their own display applications, adds complexity and time to the interpretation of the data. Silos of data often lead to duplication of servers and disparate IT departments. One of the desired outcomes for smart buildings is the ability to unify all data from multiple sources and present them in a single holistic view.

SPOG is the term that is used for this approach, which allows data to be displayed in understandable and actionable ways. It is the starting point where operators can get a sense of the greater picture of what is occurring in their building.

There are two types of SPOG views:

- **Dashboards**: Data is displayed in a dashboard format, typically using a graphic interface, making it easy to interpret the data and generate custom reports
- **SPOG UI**: This unified service allows individuals to access collaboration, communication, and experience tools through a single view, eliminating the need to switch between applications

There are the following requirements to ensure the data presented is useful:

- **Monitoring accessibility**: All of the various building systems need to have IoT devices connected to them to monitor and collect data as valid sources of information to make critical business decisions.
- **Integration**: The purpose of SPOG is to have a single platform to view all systems at once; therefore, it must be able to integrate with the existing building systems. Replacing existing systems is often not feasible; therefore, proprietary and open source system data must be able to be integrated.
- **Interpretation**: The data needs to be analyzed to spot trends and develop actionable intelligence.

- **Flexibility and customization**: Dashboards need to be flexible to allow for large volumes of data customization to generate custom reports. This should also include local and remote monitoring and notification of issues via alerts when needed.

- **Automation**: Important information needs to be filtered from unimportant information automatically to keep a user informed of the critical information required to make decisions. Built-in escalation points should be made available so that others can take appropriate steps if needed, if key personnel are not available. This is typically achieved by using and learning from data collected in previous similar situations.

- **Mobility**: A best-in-class SPOG allows users to view the data away from a building to instantly take any required actions.

Many believe that the delivery of SPOG is mythical or hypothetical, as there are several challenges that must be overcome:

- **Silos**: Rouge tools developed over time, proprietary software languages, numerous DevOps and IT operations analytics tools, and disconnected data create data silos that are unable to connect and communicate with each other.

- **Legacy stacks**: Heavy investment in the past may not be easy to replace or modify until a return on investment has been achieved. Ultimately, APIs should enable these systems to work together; however, the time and costs to reach this point may be prohibitive.

- **Constantly changing systems**: It is a massive challenge to try to keep a SPOG up to date with the constant changes and updates to each of a building's individual systems.

- **Whole or none**: If one system or part of a system is down or not working properly, there is a possibility that the whole SPOG system may fail.

- **Decentralized**: Collecting and unifying data across different departments and functions with different IP support and management structures often proves to be challenging.

- **Specialization**: Is there really a need for one person to review all the data? As organizations specialize in **Development and Operations** (**DevOps**) combined, **IT Operations** (**IT Ops**), and **Operational Technology** (**OT Ops**) functions, each is focused on their specialized role and doesn't need to see the whole picture at once.

SPOG developers will need to ensure that their solution does not look like a glorified mega-dashboard. With multiple stakeholders, each will desire a personalized pane of glass, with custom views to address the workflow they're executing at one time, while others will require a different view when they're doing a different workflow. Different groups will require different views simultaneously.

Summary

A smart building ecosystem is a complex integrated system with many components. Each component on its own will not achieve the smart building goal, but when combined, we have the framework to build the ecosystem. It starts with IoT devices and sensors that monitor an environment and send data to a local or cloud computing platform for processing and viewing.

Several wired and wireless data transmission methods and communications protocols are available, and often, smart buildings combine these. Software programs and APIs are needed to process and analyze data collected by the sensors. Data management procedures are needed to control and store the massive amounts of data collected. An easy-to-use UI allows us to visualize the data.

While this chapter focused on the smart building ecosystem, the next chapter will demonstrate how to pull these components together to develop a smart building application. Several actual use cases will be reviewed along with common challenges.

7
Smart Building Architecture and Use Cases

Smart buildings today are not designed from the top down; rather, they are assembled from the bottom up. This is because IoT smart building systems are independently designed and implemented separately from each other. To achieve true smart building benefits, a top-down approach is needed to ensure all IoT systems work together.

In this chapter, we will begin by learning about what components comprise smart building architecture. We will quickly discover the challenges in developing the architecture and why some believe that it may never truly be implemented. Most initiatives are currently centered around solving specific issues, such as **indoor air quality** (**IAQ**), smart restrooms, and energy efficiency. One exception, however, is the Sinclair Hotel PoE building transformation, which will be discussed in this chapter. We will review use cases around each of the preceding issues that smart building architecture aims to solve.

In this chapter, we're going to cover the following main topics:

- The importance of having a smart building architecture to develop smart building solutions
- What are the challenges of smart building architecture?
- Use case – Tapa Inc. and Lincoln Property Company deliver IAQ monitoring
- Use case – Kimberly-Clark smart IoT restrooms
- Use case – how the Sinclair Hotel PoE solution is solving the smart building architecture challenge
- Use case – energy reduction at an Australian mall
- The importance of building cybersecurity and privacy policies into a solution

Smart building architecture to pull components together

Existing buildings, and even buildings under design and construction today, consist of independently designed and implemented building systems. The HVAC system is separate from the security system, the lighting system is completely separate from the water system, and so on. Each of these systems is designed, deployed, and even maintained by different engineering and maintenance disciplines with little to no cross-over between trades.

Each discipline competes for capital resources to acquire the latest and greatest technology, but often, each is **value-engineered** (an approach to get to the lowest cost possible) down to the basic requirements. Because of this, we can say that smart buildings are essentially designed from the bottom up, system by system, resulting in a collection of systems trying to communicate with each other.

If we are to achieve all the benefits a truly smart building has to offer, we need to approach it from the top down to ensure all the IoT systems are integrated to work and communicate together.

Before we pull the components together, we will first need to review the smart building architecture layers. Regardless of the system and IoT applications, each system will typically have two logical layers:

- **The sensor and controller layer**: The sensor and controller layer directly connects to the physical environment. The physical environment consists of things, devices, and systems that perform the sensing, monitoring, or actuating functions.

- **The system hub layer**: This layer connects the source of intelligence and connectivity to the sensor and controller layer. The system hub is where the APIs are exposed to facilitate information exchange with the smart building applications. This specialized device has software and features that are specially designed to carry out the function of the system (that is, monitoring, security, lighting, and any other function). The hub controls the system's behavior and can provide feedback to change activity at the device level. By using a short control loop, there is minimal delay between the sensing activity and the corresponding resulting action.

While IoT sensors may appear to be smart, the smartness of smart buildings comes from the top layer through a device that manages the flow of data. These devices have names such as facility controllers, edge controllers, or IoT access network controllers. In this book, we will use the term controller.

These controllers are traditional servers or cloud-based computing platforms that contain a collection of applications connected at the bottom layer to APIs from the IoT system hubs. This controller does not deliver system-specific details, but it does integrate across systems to develop a higher-level intelligence to create the smart building.

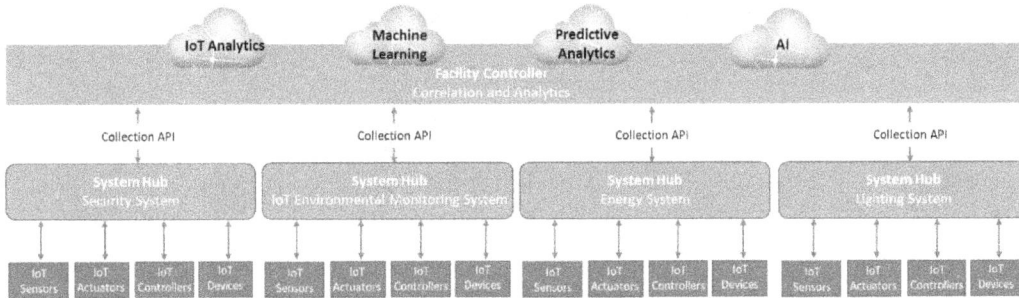

Figure 7.1 – Smart building architecture

The facility controllers consist of three cooperative elements in the smart building architecture:

- **Collection APIs**: These create links with the various system hubs that allow the controller to see events and then generate a corresponding response. While not all data from the system hubs is captured, special conditions will be extracted, and signals will be sent to the controller. These APIs will also push policies to the hubs so that they may directly respond by themselves.

- **Correlation and analytics**: Acting as the heart of the controller, this element identifies the technology options available, including machine learning and **artificial intelligence** (**AI**) technologies. With system-wide view and the use of complex event processing, simple condition reports can be distinguished from abnormal conditions.

  A systemic view of conditions includes basic items such as time, date, weather, and other conditions that can be collected from the system hubs. This information is critical in system-specific conditions such as needing to know whether the HVAC system is operating correctly. System-specific conditions require policies and rules to normalize the events.

- **Rules and policies**: System-wide conditions are tested by rules that instruct the controller to send commands to the system hubs and IoT devices in the IoT network. Policies are constraints placed upon the actions initiated by the rules determined by the controller. These can be directly input into the controllers by a building specialist who defines the rules and policies, they can be developed by AI processes, or a combination of both can be done.

  Using the security system as an example, we know that a power failure is not necessarily a security alert on its own. Combined with the loss of communications and perimeter alarms, the controllers' rules and policies may then dictate that the security department and police department are to be immediately notified.

A common architecture is needed to allow the use of common software components across most, if not all, functional layers. Smart buildings will need to accommodate an unknown wide range of sensors and devices and large quantities of system hubs. Silos will continue to be created if there are too many variations in the controller's software and buildings will not be able to adopt the latest technologies.

Smart building architecture challenges

Unfortunately, the development of a common architecture is slow, and many believe there may never be an over-arching single architecture for smart buildings. Integration capabilities are an essential part of any smart building project and many solutions currently do not offer the required integration capabilities.

Let's understand some of the challenges in building a smart building architecture:

- **Different communication protocols**: Electrical, plumbing, and mechanical systems are designed to stand alone without the thought of integration. This results in different communication protocols being used, with most of them being proprietary to the individual manufacturers. It also results in multiple computing and controlling systems, each requiring different skill sets to maintain and support.

  While each system can collect data, it is not able to share this data with other systems. An open IoT software platform is needed to pull together all the devices and systems regardless of manufacturer.

- **Lack of a smart building integration plan**: As indicated earlier, buildings are a collection of systems, and each system is managed independently. Each system is designed and engineered separately from the others with little to no thought given to sharing technologies or data.

  The HVAC engineering team may decide to implement smart thermostats using Wi-Fi mesh network connectivity with the existing BACnet protocol, while the lighting engineers may decide to deploy wireless smart lights using a separate LoRaWAN network, often because they are not even aware the other network exists.

  Rarely is there a single point where all the smart building solutions come together to meet common requirements. Typically, there is no over-arching data management plan, which means each department develops its own approach, adding to the lack of integration of systems. Departments will need to collaborate and building owners are beginning to develop smart building leadership roles and lead points of contact for IoT implementation.

- **Lack of smart building industry standards**: The building industry is built around a model where each trade has its own industry association and therefore each has its own standards. As a result, plumbing, electrical, mechanical, and technology divisions all deal with different contractors, suppliers, and integrators, each bringing their own industry's preferred technology and best practices.

  The goal is to have systems communicate with each so that, for example, when smart lights turn off due to an unoccupied room, the HVAC reacts at the same time to change the temperature. Open IoT hardware and software platforms are required to resolve this challenge.

- **Technology keeps changing**: Building systems are expensive and the return on investment needs to be realized before major changes can be made, which in most cases will take many years. IoT and smart technology are advancing so fast that often, major advances have been achieved while the system is still being designed and planned.

Smart building applications and architecture require buildings to integrate **operational technology** (**OT**) network data with enterprise **information technology** (**IT**) networks. These networks have been kept apart for cybersecurity reasons and this creates yet another challenge for smart building architecture. Over the past few years, building owners and operators have recognized that these networks need to come together, which is how the **IT/OT convergence** movement started.

IT/OT convergence

A building's IT network focuses on enterprise networking, enterprise data centers, information processing, connecting devices, and local/cloud computing. These systems support financial platforms, email servers, human resource applications, and other data center and cloud applications. Simply stated, IT systems manage data.

The OT network focuses on the hardware and software that monitors and manages the building's controls, devices, processes, and infrastructure. OT devices manage the physical world, such as the HVAC, lighting, and security systems.

Historically, IT and OT have been separated and have not shared domains, information/data, or controls. Different skills are required to design, install, and maintain IT and OT domains and the only common denominator is that both domains use IP-connected equipment. Security concerns are another driver to keep the two networks separated.

Operating two separate networks is not cost-effective and can increase the chances of a system breach due to increased access points. Smart building applications require that IT and OT are integrated to connect the physical world and the data world. IT/OT convergence is the management of the data exchange between the traditionally separated building services and the enterprise side of the operations. Achieving smart building goals requires the blending of the IT and OT networks, systems, teams, and processes while maintaining proper cybersecurity protocols.

IT and OT can be combined and consolidated on one network with the proper attention given to security. The logical way to do this is by implementing network segmentation to separate application traffic on the IP networks. **Virtual LANs** (**VLANs**) separate broadcast domains and IP subnets in the network, creating logical segmentation. **Access Control Lists** (**ACLs**) are network filters that control incoming and outgoing traffic. They are basically the set of rules that we discussed earlier in the previous section of this chapter.

For example, the HVAC system, its supporting sensors, and its management platform can be located in a separate VLAN and IP subnet. This will allow for unobstructed communication between the devices. Inter-VLAN communications can be restricted between the HVAC VLAN and other systems, such as the security system VLAN. This will ensure that a breach in the HVAC system will be isolated to that system alone.

The smart building architecture is challenging to achieve with the current system-by-system approach. Later, in *Chapter 14, Smart Buildings Lead to Smart Cities*, I'll expand upon an alternative approach with the evolution of smart buildings to unified buildings. For now, and for the foreseeable future, smart buildings are a collection of connected smart systems. Let's look at some examples of smart building use cases at the system level. The first use case is an application of the IAQ monitoring system we outlined in *Chapter 3, First Responders and Building Safety*.

The remainder of this chapter will explore a number of different use cases from IAQ monitoring and smart restrooms, which add sensors to the IT and OT network, to the use of **Power over Ethernet (PoE)** in the Sinclair Hotel, which completely integrates the networks into one. The final use case will explore the use of AI for energy management.

Use case – IAQ monitoring

"The most common way COVID-19 is transmitted from one person to another is through tiny airborne particles of the virus hanging in indoor air for minutes or hours after an infected person has been there," said *Alondra Nelson*, head of the White House Office of Science and Air. As well as COVID-19, nearly every virus is transmitted through the air, and building owners and operators are now realizing that IAQ monitoring is a long-term investment.

A Facilities Net article by Dave Lubach published on May 2 2022 titled *Occupants Unhappy with Return-to-Work Health and Safety Measures: Survey* indicated *"employees are still concerned with returning to buildings that are free of germ-causing agents on surfaces and include less-than-clean air."*

As workers return to their offices post-pandemic, building owners and operators are faced with the challenge of assuring these occupants that their buildings are clean and safe. In addition to demanding healthy buildings, occupants also want more control of their overall environment, and IoT technology solutions are delivering these combined capabilities. It's not enough to just tell the occupants that mitigation efforts have been taken; they want to see demonstrated real-time results.

Lincoln Property Company (LPC) in Washington DC partnered with Tapa Inc. to implement Tapa's IAQ monitoring solution to assure occupants that their IAQ met or exceeded the industry-recommended levels. LPC chose its 101 Constitution Ave location in Washington DC to launch this program. This is a 10-story building with 511,338 square feet of office space and it is the closest building to the US Capitol building.

This is a textbook example of using IoT to make a building smarter. In this case, Tapa first installed their Tapa edge server to provide the computing power. The server was connected to the building's existing BACnet communication network, allowing the server to communicate with other connected IoT devices and controllers. It also provided a connection to the building's HVAC system to collect air flow information to determine the air exchange rate per hour.

Next, air quality sensors were installed in the common areas on every floor. Additional sensors would be added over time and based on tenant requests for their space. Each sensor measured temperature, humidity, **carbon dioxide** (CO_2), particulate matter (PM, PM2.5, and PM10), and **Total Volatile Organic Compounds (TVOCs)**. Industry and government guidelines for recommended levels exist for each of these environmental elements measured. Complex analytics combined these elements to deliver an IAQ score on a scale from 1 to 100, with 100 being great air quality.

Once the IAQ sensors were installed and powered up, they needed a way to communicate and transfer data back to the Tapa edge server. Given the unique nature of this building and to meet the client's aesthetic requirements, two connection methods were used. In some locations, the sensors were directly connected to the edge server via Ethernet cables, and in other locations, the sensors were connected using a private Wi-Fi mesh network installed by Tapa.

Once the sensors were turned on, they began communicating with the edge server and providing real-time data. These sensors required a few days to self-calibrate but the readings already provided valuable air quality data. The analytic software on the edge server calculated IAQ scores for each sensor, each floor, and the entire building. Early IAQ scores were above 97, indicating the air quality in the building was very good.

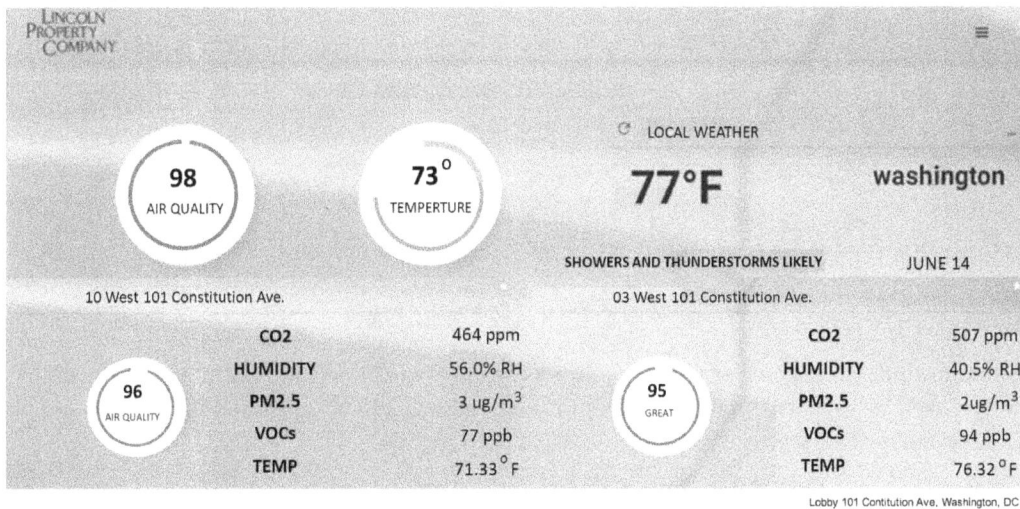

Figure 7.2 – LPC 101 Constitution Ave IAQ readings

After the sensors stabilized over a few days, the readings improved by 1 point to a building average IAQ score of 98. *Figure 7.2* shows the building's total score of 98 and 2 individual sensor readings, one from the 10th floor west side and the other from the 3rd floor west side of the building.

The next step was to communicate these results to the building's occupants. Three methods of communication were selected. The first method was to place a small desktop tablet on the entrance security desk to show incoming visitors and occupants the building's total score. Tablets were also placed in the building's engineering department so that they could monitor scores and take any action if required.

Second, a widget was built and added to the entrance lobby's digital display so that everyone could see the real-time results. *Figure 7.3* shows the information displayed on the digital sign.

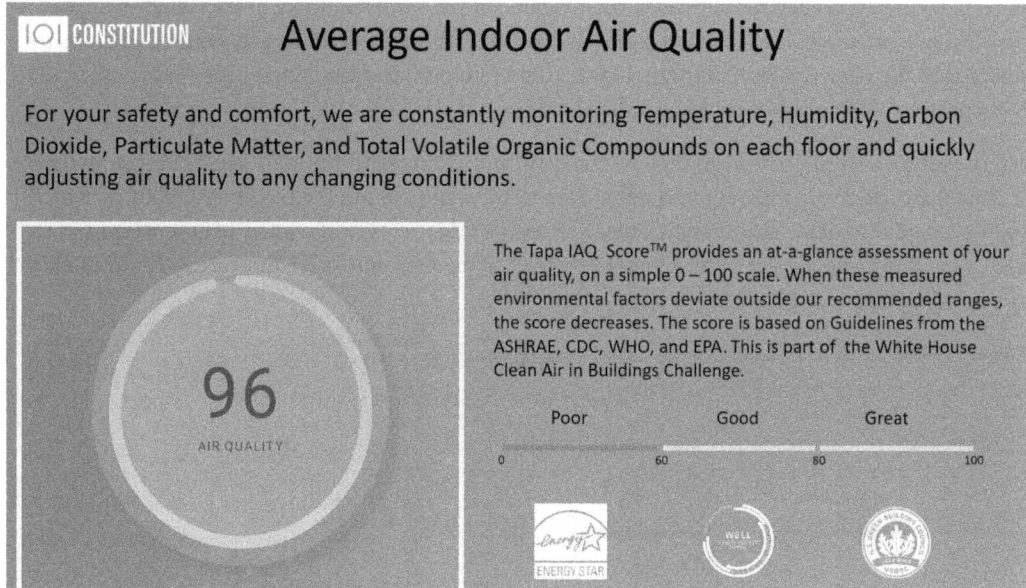

Figure 7.3 – Lobby digital sign displaying IAQ information

The third method was to provide building occupants access to a web widget to deliver the information directly to their smartphones. In this case, they were only given access to the entire building's collective IAQ score and the IAQ score for common areas and their space.

The building engineers, on the other hand, had access to a detailed breakdown minute by minute. If they noticed a reading falling below the recommended level, they could immediately take mitigation actions, such as increasing or decreasing the outside air flow, turning on dehumidifiers, and activating ionization systems.

Unexpected benefits

The successful implementation of the IoT sensors, edge server, communication channels, and analytics software delivered the desired results and provided real-time IAQ readings. Occupants felt comfortable knowing the air in the building exceeded the recommended requirements.

After the system was up and running for about 2 weeks, the building engineers began to review trending analytics data. While conducting this review, they discovered that the IAQ scores on the east-side floors were dropping after-hours but returning to the proper range later in the evening. The 55% to 80% humidity levels in one of the risers were impacting the IAQ score as these were well above the IAQ threshold of 40% to 50%.

Upon further review, it was determined that a sequencing issue was causing an outside air fan to continue to run after-hours, filling the air handling rooms with warm and moist air. This warm air then had to be cooled, creating an unnecessary cost. Had this sequencing continued, mold would have built up, causing major issues and costs to clean. With simple adjustments made, IAQ monitoring not only helped save energy costs but also prevented a potential mold build-up.

The building engineers continued to review the trend data and discovered that CO_2 levels never came close to the threshold of 600 ppm. The engineers were able to reduce the amount of fresh air that had to be heated or cooled and therefore saved additional energy costs. In the end, the smart building benefits within the first month alone paid for the cost of the IoT IAQ solution.

Use case – smart IoT-connected restrooms

Auto-flush toilets, touch-free paper towel dispensers, and auto-on-off water faucet restroom solutions have been in place for years. The pandemic has greatly increased hygiene expectations and the proliferation of IoT devices has increased the range of touchless devices and smart data-driven solutions available. On average, individuals will visit the restroom three to four times per day, touching many of the commonly used surfaces, greatly increasing the chances of virus transmission.

According to a recent Kimberly-Clark Professional survey, 80% of workers feel that the restroom is one of the most important areas requiring better hygiene levels. Mark Caskey, CEO of JLL's EMEA Corporate Solutions business, stated, "*Employee health and workplace hygiene are, without a doubt, the top considerations for companies re-entering the workplace,*" and the **Centers for Disease Control and Prevention (CDC)** stated, "*hand washing is a simple thing to do and it's the best way to prevent infection and illness.*"

Post-pandemic, buildings have made changes to their restroom routines by increasing the number of times they clean the restroom, along with developing more thorough cleaning and disinfecting routines.

The Hygiene Behavior Consumer Study by Kimberly-Clark Professional, from May 2020, found that odors, unflushed and/or dirty toilets, and empty dispensers impacted cleanliness perceptions, largely in office, educational, and other high-traffic buildings. 73% of tenants said a bad restroom equates to poor management and 60% said that an unhygienic restroom lowers their opinion of the building and chances of resigning their lease.

Kimberly-Clark has developed a smart restroom solution that they've branded Onvation®. Its system monitors, measures, and analyzes information collected from connected IoT devices and smart software.

Onvation® works by utilizing the following:

- **Smart sensing**: Smart IoT dispensers, fixtures, and sensors collect real-time device status and usage data. IoT-enabled sensing items include soap dispensers, paper towel dispensers, toilet paper dispensers, air fresheners, hands-free faucets and toilet flushes, automatic doors, and trash bin level sensors.

  The network provides low-stock or out-of-stock notifications, low or depleted battery status, dispenser jam or clog notifications, and spill/overflow notifications. Connected plumbing, work order systems, digital signage, and occupancy tracking can also be added to the system if desired.

- **Connection**: A local gateway is installed to securely connect and communicate with the smart IoT devices and send the data to a cloud-based computing platform called the Onvation® cloud. A third-party gateway connects to the Microsoft Azure cloud storage and computing solution. This information is continuously uploaded using either cellular or Wi-Fi connections.

- **Actionable insights**: Onvation® receives the data and then aggregates and analyzes it to develop actionable responses. Real-time notifications can reduce waste and eliminate runouts. Proactive maintenance can prevent issues and analytics can lead to more efficient staff deployment and improved operational efficiency.

- **Mobile apps**: Connection with a mobile app maintains constant communication with the cleaning staff to replenish products and clean facilities proactively.

Kimberly-Clark's smart restroom solution saw the following achievements:

- 75% reduction in complaints and work orders
- A reduction in consumable waste of up to 80%
- A 90% reduction in time spent checking dispensers
- Analytics data produced for product consumption, activity history, restroom traffic, and device readiness

Smarter restrooms allow building occupants to begin to see and feel the smart building experience. IoT technology solutions deliver real-time information to improve visibility, intelligence, and control to deploy cleaning crews when and where needed.

Use case – PoE at the Sinclair Hotel

I toured the Sinclair Hotel and met with Sinclair Holdings' president and developer, Farukh Aslam, while it was transforming to become one of the smartest buildings in the world. Now open, this massive undertaking is the first of its kind using PoE to power everything, and it is the first building in the world to power itself.

Built in 1929 in Fort Worth, Texas, US, this 17-story building originally served as the corporate headquarters for Sinclair Oil and later transformed into the Sinclair Hotel with 164 rooms. Located in the Art Deco district and needing a major renovation, the developers and owners of this 90-year-old building wanted to retain its unique facade and art deco interior, while making it ecologically sound.

Farukh and his team installed PoE technology and redesigned traditional equipment to run on low voltage to power lights, shades, curtains, windows, smart mirrors, hairdryers, minibars, TVs, AC units, exercise equipment, door locks, and everything else that required power using a standard Ethernet cable. He stumbled upon this solution while renovating another building and running into LED light-dimming issues. In his search for answers, he found Cisco had the technology to deliver a PoE lighting system that was capable of dimming. Farukh thought to himself, if they can do it for lighting, why not for everything else?

2,000 lights and amenities and 7,000 ports are powered with PoE technology, and each has an IP address on the computer network. If any PoE device or light goes out anywhere in the building, the building manager is immediately alerted with a notification on their smartphone. Management can turn lights on and off from any device or smartphone, requiring simply an internet connection.

The building's energy consumption was reduced by up to 40% and 23,000 USD were saved during the retrofit since they did not have to hire electricians or even need to obtain electrical permits. Digital electricity is used everywhere and connected via Ethernet cables. It is AC power that is converted into DC power and transferred as packets of energy like a low-voltage system.

An old, bulky diesel generator was replaced with the world's first UL 924 lithium-ion battery system, which Farukh discovered at an LG office in Korea. The environmentally friendly battery system can provide 3 hours of power during an outage and requires much less space, freeing up valuable useable space in the hotel.

Vendors were asked to create products that could run on PoE so that the entire building could use this technology. New motors were designed for the window shades and LG delivered LG VRF air-conditioning units, LG PoE OLED TVs, and LG PoE wallpaper TVs. Savvy smart mirrors were installed that provide guests with special lighting selections and built-in apps that display the morning weather and news directly on the mirror.

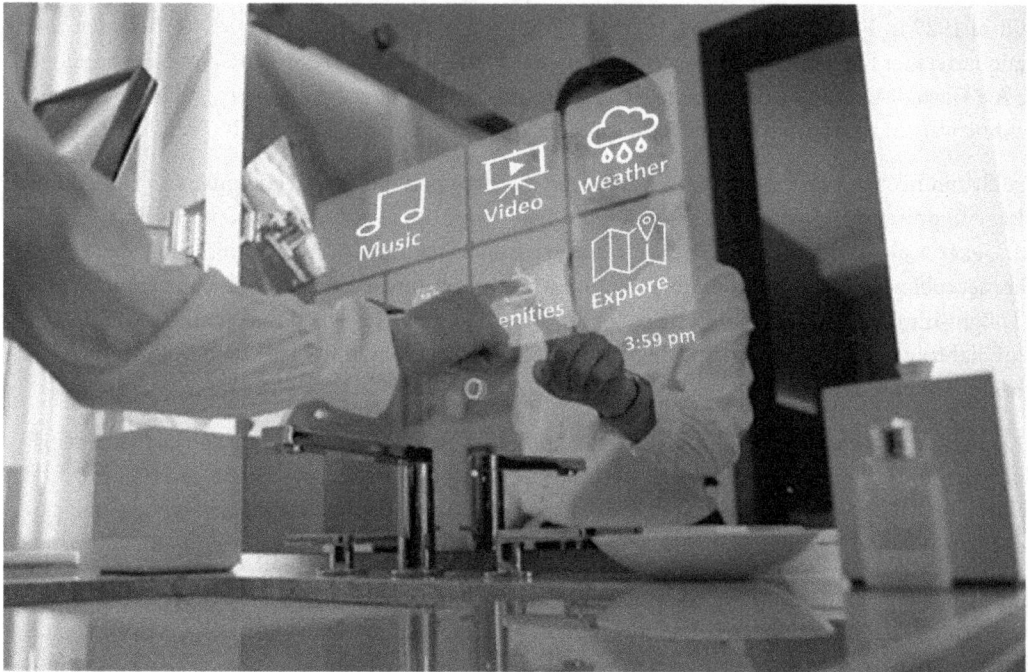

Figure 7.4 – Smart mirror technology (Image credit: https://electricmirror.com/savvy/)

The minibar was converted to use PoE power and the organization Italian Technogym created special exercise equipment that powers the hotel while guests are doing their workouts. Marriott reward members can earn reward points for completing a 20-minute cardio workout that supplies power to the building, as well as a carbon footprint reduction certificate at checkout.

Creative lighting has been installed on the exterior and rooftop of the building and each evening, a different lighting sequence is displayed, creating a unique city attraction. Often, children staying at the hotel are given control of this lighting system for fun. Wireless access points are placed throughout the entire building for seamless guest internet access.

Kohler developed Kohler Konnect™, a digital shower that uses voice-enabled technology to deliver unique showering experiences that are designed by each guest. An invigorating shower coupled with innovative lighting selections delivers guests the ultimate shower experience. SinkTech IoT sinks regulate water temperature and soap and sanitizer levels in the hotel's restaurants.

How it works

Ethernet cables replaced the old electrical wires throughout the entire building, along with an 18/2 wire and speaker cable, which is used for the homerun cable to each guest room. Cisco Catalyst Digital Building Series switches are tucked away in each guestroom closet and power is supplied by

VoltServer's Digital Electricity. An Intel PoE NUC IoT gateway supports edge computing, controllers, data aggregation, power management, and the Intel Unite hub. The Intel Unite hub provides wireless displays throughout the building and is the platform used to power and manage the building's meeting spaces.

Intel provided end-to-end software that delivers smart features, such as the reservation system, networking infrastructure, point of sale, guest services, mobile keys, and back-office services. As guests enter the hotel's bars, their smartphones receive a push message welcoming them to the bar and offering them a drink discount or informing them of special menu items. Guests' orders are remembered so that the next time they enter the bar, additional special menu items may be offered based on their history and preferences.

Bluetooth sensors and mesh networks are installed in each guestroom, which tell management and housekeeping that the room is occupied. The temperature is automatically adjusted up when the room is unoccupied, and when guests reenter the room, the temperature is automatically adjusted back. This same technology is used for people counting in conference rooms to optimize HVAC levels.

Cisco provided their smart Wi-Fi cloud networking solution Meraki, complete with SAS data analytics, to deliver personalized guest messaging and location-based analytics.

Numerous other IoT and PoE technology solutions have been installed and massive amounts of data are being developed everywhere. The Sinclair has placed an Intel NUC quad-core processor with 1 terabyte of storage and 32 gigabyte RAM systems throughout the building to process data, run analytics, and provide the computing power to eliminate latency so that when a guest walks into their room, the light turns on instantly.

Solving for the smart building architecture

Farukh and his team are now focusing on solving the top-down IoT architecture we outlined at the beginning of this chapter. While the IoT and PoE infrastructure has been installed, each system is still managed and controlled by its individual siloed controllers with its own API. For example, the LG Smart AC has its own built-in state-of-the-art energy management system and the lighting system has its own state-of-the-art light management system.

Each device now has an IP address that can be accessed and controlled individually if desired. IP address connectivity eliminates the use of BACnet-based systems because it can connect to every device directly, and the GUI-based interactive system allows for easy communication.

Working with Intel, LG, Cisco, VoltServer, and SaaS partners, Sinclair Holdings is developing a completely new IP-based **building management system (BMS)**. The goal is to create a system that can connect, collect, and process data from every device and system in the building on a single platform. Digital twins (to be discussed in detail in *Chapter 8, Digital Twins – a Virtual Representation*) and AI are being used to run analytics.

Use case – energy reduction at an Australian shopping mall

A leading investor in the Asia-Pacific region has built their portfolio of buildings with a heavy focus on environmental, ecological, and sustainable goals. Their latest project involved finding ways to reduce carbon emissions and eliminate unnecessary operating costs at their 215,800-square-foot shopping mall in Australia. Of course, they also wanted to improve the visitor experience.

BrainBox AI, a Montreal-based company, was brought in to use their predictive and self-adapting AI algorithms to optimize the HVAC system. They developed a custom driver for the Tridium (a leading IoT software provider) Niagara Framework to connect to the HVAC system. Using custom-curated algorithms, AI, and cloud computing, they were able to pull down a few weeks of actual data from the system.

After an AI learning period, building analysis, and data mapping activities, BrainBox AI developed a unique strategy for the building. They combined the data they extracted from the HVAC system with external weather and tariff structure data to drive reductions in runtime for the building's assets and equipment, along with reducing the HVAC system's energy consumption.

The result was a saving of 21% in electricity costs in 5 months by reducing 29,855 kWh, translating to 4,776.80 AUD (roughly 3,303.19 USD). The mall avoided 32 metric tons of CO2eq and achieved a 25.25% reduction in supply fan runtime, along with a 43.5% reduction in compressor runtime.

Cybersecurity for smart buildings

As buildings become more connected, they are more vulnerable to cyber-attacks. Cyber-attacks continue to rise year on year and infrastructure is a common target. An attack can have devasting consequences, from exposing sensitive information and shutting down or disrupting critical building services to compromising occupant safety. The estimated average costs of a data breach are 4.24 million USD. Ransomware attacks can create safety issues as hackers disable locks, alarms, elevators, and other critical building controls to prevent people from exiting or entering the building.

Buildings are an attractive target since they are often not well protected and have critical control systems. They often contain other non-building-related IT databases and systems. In an earlier chapter, we highlighted that the Target store data breach a few years back resulted from hackers entering their confidential database through the HVAC system vendor's portal.

The building's reputation can be greatly harmed and compliance requirements such as the **General Data Protection Regulation (GDPR)** and the **Health Insurance Portability and Accountability Act (HIPAA)** can be significantly compromised.

Human error is a leading breach cause as individuals leave systems open, download malicious software accidentally, and do not follow password protocols. Running outdated operating software opens access points for hackers and is more vulnerable to malware. Smart building automation systems are typically not encrypted and often lack security protocols.

By hardening the building's systems and increasing security protocols, hackers will have a much more difficult time accessing the building's systems. This can be done with the following:

- **Security system**: A robust system includes cameras, firewalls, intrusion detection, and incident response plans:

    - Security cameras provide monitoring, determent, and historical information capabilities for post-incident reviews

    - Properly configured firewalls block unauthorized traffic from accessing the network

    - Intrusion detection identifies suspicious network activity and provides immediate alerts

    - Having an incident response plan in place helps the team to deal with security incidents quickly and effectively

- **Secure protocols**: Secure protocols are available and should be used for communication between all devices.

- **Device security measures**: Access control and encryption should be deployed as security measures at the device level. By controlling who has access to the building, unauthorized individuals can be kept out of sensitive areas. This will make it harder to access data and keep them out of the network. Encryption can protect data at rest and data in transit.

NIST cybersecurity best practices

The **National Institute of Standards and Technology** (**NIST**) has developed and disseminated IT/OT standards and best practices to help smart buildings address cybersecurity threats:

- **Identify**: Knowing what you are protecting is critical and NIST recommends developing a list of all equipment, computers, laptops, smartphones, IoT devices, servers, sensors, tablets, point-of-sale devices, and all other connected mechanical equipment. Everything on the network needs to be accounted for. Low-impact scans can help identify rogue devices, which will need to be eliminated.

- **Protect**: Segment the network by storing sensitive data in separate secure network locations. Assign roles to everyone, including vendors, and limit vendor access where possible. Perform vendor audits more frequently. Document the cybersecurity framework and ensure data at rest or in transit is encrypted.

- **Detect**: Security software should be kept up to date and backups should be performed routinely. Monitor network traffic and review traffic patterns, often looking for items that appear strange.

- **Respond**: It's not a matter of if but when an attack will happen. Therefore, written response plans are critical to have in place to document recovery and alternative processes to keep the building running. Notification of data breaches should be communicated to all stakeholders who might be at risk.

- **Recovery**: A formal detailed and documented recovery plan should be in place and well communicated. Having backed-up data on different machines is critical for recovery but remember that using a third-party vendor to manage backups may cause delays.

NIST reminds us that this cybersecurity framework is a mindset that should be part of all owners' and operators' decision-making process and to protect their occupants and assets, they must make cybersecurity a top priority. With IT/OT convergence, a shared end-to-end responsibility, policies, procedures, technologies, and governance combined, a common strategy can be delivered.

Smart building IoT and privacy

IoT devices deliver valuable insights into what's going on in and around the building and are a key element in delivering healthy, safe, and secure buildings. Unfortunately, building owners and operators lack the technical knowledge to secure their buildings and they put their occupants' information at risk. Overbearing surveillance and data collection policies can create privacy concerns for the building's visitors and occupants.

Privacy and cybersecurity are often discussed together and even sometimes confused with each other. We simply differentiate them with the following summary: privacy is how the data is used, and cybersecurity is how the data is protected.

Cameras, facial recognition devices, occupancy sensors, license plate scanners, microphones, and many other IoT devices provide valuable smart building data required to control and manage systems. While most buildings will collect this data and autonomize it, occupants remain concerned about how the data is used, stored, and protected.

Privacy concerns vary greatly across the world based on cultural acceptance, industry type, and government policies, making it difficult to recommend a single solution or policy. Given this, we suggest a few things that smart building operators should do to address privacy concerns:

- **Don't go overboard**: Just because you can deploy as many surveillance cameras as you want, doesn't mean that you should. Occupants will understand the need for surveillance at entry/exit points, but other areas need to be off-limits. This includes areas where sensitive business information could be viewed or heard, such as conference rooms and private offices.

- **Be upfront**: Occupants may feel more comfortable if they are made aware that data is being collected not for individual tracking and surveillance purposes but to improve heating, cooling, and other environmental comfort factors. Tell them what personal data is being collected, why it is collected, and how it is being used.

- **Document policy**: A data collection and usage policy should be created, documented, and distributed. It should highlight what data is collected, how the data is used, and who owns the data. Information should be provided on how the data is anonymized or abstracted. The document should also indicate whether the occupant can choose to opt out of any collection practices. The **European Union's (EU's)** GDPR is becoming the de facto policy to protect individuals' privacy rights.

- **Physical controls**: Physical security is important as well, such as limiting access to certain offices and areas of a building where sensitive information may be stored. With the post-pandemic increase in hybrid work and school environments, individuals feel more comfortable using their own devices for work. This can be risky and lead to breaches; therefore **bring-your-own-device (BYOD)** policies should be developed and enforced.

Privacy concerns will continue to shift as building owners provide informative details that address occupants' concerns. Post-pandemic, we are seeing a shift in concerns and occupants understand that many of the IoT solutions implemented directly impact their health and safety.

Summary

Smart building architecture is needed to connect all the various building systems in one platform. While there are many challenges to implementing this, proper planning and attention to a top-down approach can resolve these. We provided four IoT smart building use cases, each demonstrating the application of the IoT device technologies, communication protocols, connectivity, computing, and data management components discussed in the previous chapters, now all pulled together to deliver real results. Connecting all these building systems requires a major focus on cybersecurity from the start.

This chapter demonstrated how to pull these components together to develop smart building applications. In the next chapter, we will explain digital virtual representation and digital twins.

8

Digital Twins – a Virtual Representation

One of the fastest-growing smart building technologies today is the creation of a virtual representation of buildings, equipment, and systems using a technology called digital twins. Virtual replicas of a building are being used in the building construction process and for building operations, maintenance, and management of day-to-day processes.

In this chapter, we will learn about smart building digital twins and digital representation and how the information required to build the digital twin begins with the data collected, monitored, and managed by the IoT devices. Digital twins provide the dynamics of how the internet of things devices operate throughout their life cycle.

In this chapter, we're going to cover the following main topics:

- Defining smart building digital twins and providing a brief history, along with many building application examples

- Explaining the different types of digital twins in a smart building

- Discovering the layers and components that form a digital twin

- Learning how to create a digital twin

- Examining the importance of digital twins and public safety, with first responders' real-time access to critical data

- Exploring the significant contribution digital twins deliver to hospitals and how a large home improvement box store is implementing digital representation to improve the shopping experience

- Understanding the challenges smart buildings are facing trying to implement digital twin technology

Smart building digital twin defined

As buildings continue to implement IoT projects, so too are they implementing digital twin solutions in large numbers. If they are not currently implementing digital twins, they are in the process of planning to establish a digital virtualization strategy. Quite simply, a **digital twin** is a virtual replica of a physical building, its systems, assets, people, devices, places, and processes. It is a contextual model not just of the physical environment but also of the entire organization, operations, and processes.

Digital twins are used to monitor energy consumption, waste management, air quality, fluid leaks, HVAC, and more. They provide a deeper understanding of occupancy patterns and provide a virtual experience of the space before expensive capital improvements are made. For businesses where it is critical to deliver a consistent customer experience across all locations, digital twins provide a schematic of how to deploy devices, systems, and processes consistently.

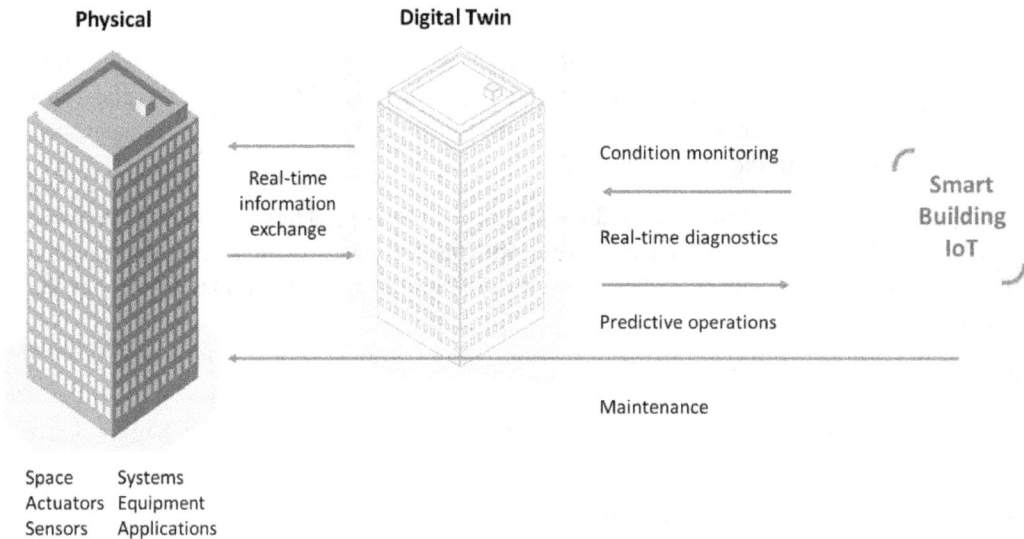

Figure 8.1 – A digital twin overview

As displayed in *Figure 8.1*, real-time interaction data from people, processes, and connected devices are brought together with data from subsystems and other data sources. Owners and operators gain better insights into a building system's performance to quickly identify issues, optimize efficiency, improve space utilization, and enhance safety. It also acts as a simulation tool to determine the potential impact of implementing new smart building technologies.

Each of the building's physical items, such as HVAC systems, elevators, lighting systems, door locks, motion sensors, surveillance cameras, and any other physical system, can be replicated individually or collectively. From here, building operators can run simulations to test outcomes they need or want to happen. Simulations such as a fire breaking out can be conducted virtually to not impact the assets or the building itself.

History and use in smart buildings

John Vickers of NASA first used the term *digital twin* in the 1960s when NASA used twinning to physically duplicate their space system on the ground to mimic Apollo 13 conditions and scenarios. As information technology evolved and more data became accessible, the manufacturing sector adopted digital twins in the 2000s.

Early use of digital twins replicated a single asset or system with a focus on the detection of issues, predictive maintenance, and asset reliability. As they have grown to become more complex, digital twins now focus on systems of assets and entire buildings to understand how people, assets, and workflows work together to manage an entire building throughout its life cycle. They bring together IoT sensors, third-party data, IT, and OT systems to deliver a complete digital replica of the building.

Smart building digital twins are not **Building Information Models (BIMs)** or 3D visualizations of a building. BIMs are used by architects and engineers throughout the design-build process, and they focus on the physical space and can be provided data from a digital twin. Expensive 3D visualizations can be developed from digital twins' insights; however, they do not provide insights into the people and processes to manage the operations. Typically, the digital twin will provide insights for more cost-effective problem-solving solutions, such as 2D floor plans and dashboards.

A smart building digital twin use example is the simulation of airflow through a space before it is built to determine the air-exchange rates and space temperature impact from stationary and dynamic objects, such as partitions, furniture, and people. Another example is in the design of evacuation scenarios to determine the best routes and which doors might need to be made wider. Managers use it for return-to-office scenario planning and continuous building monitoring and asset tracking.

MarketsandMarkets Research forecasted in September 2020 that the global digital twin market will grow from $3.1 billion in 2020 to $48.2 billion in 2026 (CAGR of 58%). They suggested this will largely be driven in the healthcare and pharmaceutical industries because of COVID-19. Climate change is another driver, as building owners and operators look for ways to improve efficiency and reduce the carbon footprint. Mind Commerce in March 2021 predicted that 90% of all IoT platforms by 2026 will contain some form of digital twin technology.

At a high level, smart building digital twins provide a contextual model of the building and its operations to do the following:

- Provide spatial awareness
- Deliver intelligent recommendations
- Provide root cause analyses
- Facilitate self-tuning

- Provide data for predictive maintenance

- Support scenario planning and what-if analyses through simulations

The digital twin continues to get smarter as more and more data is added and can be used at every stage, including design, build, commissioning, operations, and maintenance. It benefits all the building stakeholders, from owners, operators, and managers to occupants and tenant employees, with real-time data, workflows, and behaviors to optimize performance.

Digital twins in a well-planned smart building can be used to deliver the following:

- **Building insights**: Digital twins provide better insights to support strategic decisions on how to operate a building. Detailed information can be analyzed and simulated for any building system.

- **Energy management**: Energy efficiency and sustainability goals can be achieved and improved by finding patterns in the supply and use-consumption data. HVAC maintenance and operations data is used to drive energy demand and management. Real-time space utilization data can activate responses to energy/water transmission, generation, storage, maintenance, and management.

- **Predictive maintenance**: Strategically placed IoT sensors and devices coupled with machine learning and artificial intelligence algorithms predict patterns to drive maintenance operations, to deliver optimal repairs and service. Parts can be ordered well in advance and technicians can be scheduled to perform the work.

- **Return to office**: Safety and wellness programs, contact tracing, space reservations, and workplace experiences can be modeled and adjusted accordingly.

- **Workplace health**: Environmental measurement and management scenario planning, such as indoor air quality monitoring and management.

- **Space utilization**: Occupancy sensors combined with artificial intelligence can determine how space is being used and spot space that is underutilized. Access control and temperature management can be integrated to derive clever solutions.

- **Navigation**: Easy-to-use digital navigation systems can guide visitors to the correct space.

- **Automation**: Repetitive tasks can be learned and automated.

By using smart building digital twin technology, building owners and operators can bring together unconnected systems to gain new insights, monitor processes remotely, and optimize workflows.

Monetization opportunities

While cost optimization is a primary driver for the use of digital twins, value-added services and operational efficiencies are key to monetization. Let's take a more detailed look at some digital twin use examples in the building operations and management area:

- **Monitoring**: Dynamic or intelligent condition monitoring is not new; however, the existing solutions are inefficient, produce inaccurate data, and do not deliver data fast enough for a quick response during fast-changing events. That's because these systems typically use historical and static data that lags behind the current data and does not represent the current state.

Digital twin models work with data that is being collected in real time from sensors, detectors, fixtures, and other systems located in a building. Because the digital twin is constantly fed real-time information, it can mirror exact conditions on site and respond with immediate actions, such as when a faulty sensor or leaky valve is detected.

- **Occupant management**: The digital twin is the digital portal to provide access to tenant services for the occupants. Building managers can access reservations, conference room bookings, and parking space utilization and use this information for department planning, workforce interactions, and logistic improvements.

- **Smart parking and EV charging**: Garage cameras, parking space sensors, and EV charging stations are connected to the digital twin to increase safety and to measure parking space utilization. From here, space-sharing scenarios, parking space reservations, and dynamic pricing can deliver monetization opportunities.

- **Asset maintenance and traceability**: The digital twin enables users to interact directly with on-site equipment to track location and review maintenance records. This data is used for predictive and preventative maintenance routines. Blockchain is used to track and inspect supply chain materials and building parts.

- **Data centers**: Advanced data centers require server rooms and storage racks to have special cooling and other environmental conditions that can be monitored and maintained through digital twins.

Energy efficiency and sustainability gains from digital twins include the following:

- **Electric demand optimization**: Virtual power plant actuation and monitoring systems use advanced algorithms for the optimization of a constantly changing energy demand. The digital twin allows optimization actions to be made visible and accessed retroactively, providing transparency and detailed records.

- **Energy efficiency**: Energy consumption per room can be viewed using digital twin virtual sub-metering analysis. Advice is provided for operational process improvements, and utility companies offer incentives for energy optimization and flexibility.

- **Sustainability and ESG reporting**: Building certification and well-being programs require strict recordkeeping for proof of compliance necessary to secure funding. Digital twins are used as the system of record and to communicate updates to a building's occupants.

Predictive analytics uses digital twins to monetize data with the following methods:

- **Simulations**: Crowd flow, airflow, and indoor air quality are examples of where digital twins can run what-if analyses and simulations based on physical models.

- **Predictive analytics**: Camera systems enhanced with AI can be used to identify hazards, show people's movement, enhance facility utilization, and control crowd flow speeds. Cameras are used as sensors to collect data for the digital twin.

- **Predictive maintenance**: Maintenance is one of the leading use case examples for smart buildings and Industry 4.0, and it is estimated to produce a 20% reduction in a building's energy and maintenance costs. Digital twin and IoT technologies are driving these savings by accurately determining maintenance needs before any system failures occur.

  At the center of developing predictive maintenance algorithms is the data collected from IoT sensors. Digital twin-based software collects real-time records from every physical asset within a building and will analyze that data against historical data. From there, tests can be simulated with numerous possible fault combinations and levels of severity in a controlled environment. Intentional faults can be created, and the data is used to train the classification algorithm for that specific fault detection using 3D CAD files and augmented reality.

- **Data monetization**: Digital twins can easily replay and visualize usage patterns to help building operators tailor better services. Use and consumption data is benchmarked, analyzed, and monetized.

Digital twins' technology delivers safety and security solutions such as the following:

- **Security and monitoring**: Live data from scanners such as parking garage entrances, gun-shoot detection, and weapon detectors can be triggered. Location data and dynamic information sharing are enhanced through the use of digital twins.

- **Operational safety**: Evacuation plans can be simulated as well as real-time occupancy data driven by digital twins. Improved visibility during an emergency can reduce first responder times.

- **Real-time access monitoring**: A building operator can see the layout of a building and who is logged into each workstation and for how long.

These are only a few examples where early adopters are using digital twins in smart buildings. Widespread deployment may take years to become mainstream in the building infrastructure.

Types of digital twins

Different types of digital twins exist for predictive maintenance, based on the detailed level of analysis that is required:

- **Component twins/parts twins**: Component twins are the smallest basic unit of the digital twin or the smallest operating component. Parts twins are similar and reference less important components.

- **Asset twins**: An asset consists of two or more components working together. An asset twin studies these components' interactions and turns the performance data into actionable insights.

- **System or unit twins**: System twins pull different assets together to create a fully functioning system. From here, the interaction between assets can be reviewed and performance engagements recommended.

- **Process twins**: Moving up yet another level to the macro level, process twins review how systems work together. These systems should be synchronized to be effective or tweaks may be required to deliver precise timing.

While the biggest difference between these twins is their application, it is possible to have the different types co-exist within a process or system.

Layers and components

The smart building digital twin can be categorized into four layers:

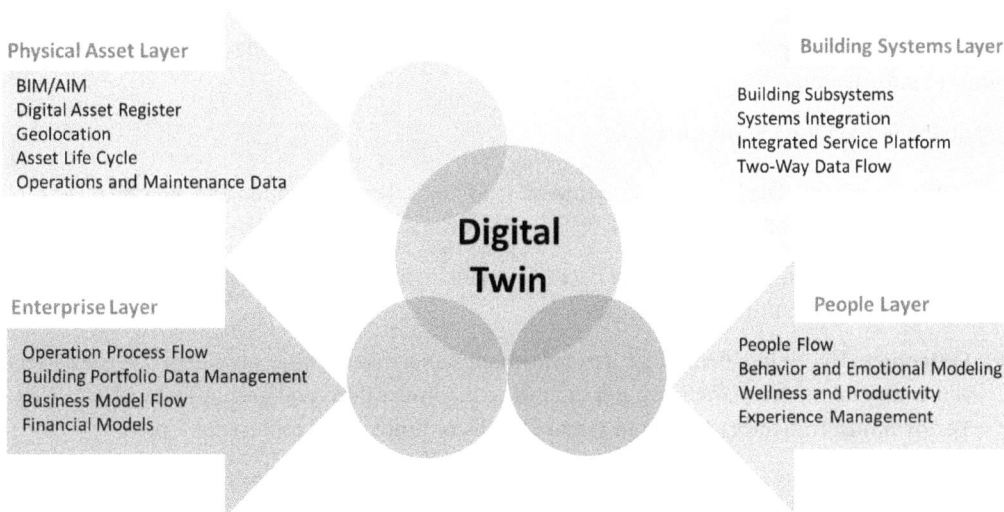

Physical Asset Layer

BIM/AIM
Digital Asset Register
Geolocation
Asset Life Cycle
Operations and Maintenance Data

Building Systems Layer

Building Subsystems
Systems Integration
Integrated Service Platform
Two-Way Data Flow

Digital Twin

Enterprise Layer

Operation Process Flow
Building Portfolio Data Management
Business Model Flow
Financial Models

People Layer

People Flow
Behavior and Emotional Modeling
Wellness and Productivity
Experience Management

Figure 8.2 – Digital twin information layers

The digital twin comprises data collected from all aspects of a building. To simplify the visualization of data, digital twin information layers represent the different categories. On the operation technology side, the physical asset layer collects data from the numerous different types of equipment, machines, and tools in the building and their connected infrastructure. All the building's systems, such as HVAC, security, fire, and access, comprise the building systems layer.

On the information technology side, a building's financial, operations, and business support systems, such as email and databases, are represented in the enterprise layer. People tracking, location, health, and well-being form the people layer. Together, these disparate systems deliver information that is typically presented through graphs, 2D charts, and illustrations of the building.

With digital twin capabilities, this information is presented using **Extended Reality** (**XR**), which includes augmented reality, virtual reality, and mixed reality:

- **Virtual Reality** (**VR**): This is a fully enclosed digital environment that delivers an artificial experience while providing no sense of real-world situations.

- **Augmented Reality** (**AR**): This is the real-world environment with digital information overlayed onto what can physically be seen. The real world remains central to the experience and is enhanced by the virtual details.

- **Mixed Reality** (**MR**): The virtual environment is mixed with the real-world environment to allow for interaction and manipulation of both.

Extended reality maximizes a true view of what is going on inside a building. Using this virtual technology, we can see through walls, look down through floors, and look up through ceilings to visualize the building to quickly find and resolve issues.

A digital twin uses high-performing databases and advanced reasoning engines to correlate massive amounts of information.

The digital twin has four key components:

- **Data**: A semantic data layer that structures data for logical representation and querying. Data is compiled from building systems, blueprints, and external data sources such as weather forecasts.

- **Reasoning**: The ability to correlate data to provide insights that can lead to outcomes using ML/AI models and/or non-linear rules.

- **Outcomes**: The ability to visualize that information in a manner that a nonexpert system user can consume, such as efficiency, net operating income, safety, and gashouse emissions **Key Performance Indicators** (**KPIs**) in visual 2D, 3D, or point cloud representation.

- **Context**: The ability to influence system and device behavior, occupant behavior, and workflows.

Recently, AR and virtualization have been applied to digital twin technology, and more autonomous and blockchain technology will be included going forward.

Creating a digital twin

A guiding principle to remember is that a digital twin is not just a replicated computer model of a building and its system; it is an extension of the environment used to manage the integrated building. With that in mind, let's look at how to create a digital twin:

- **Purpose and scope**: Start by understanding the intended purpose, such as whether will it drive navigation, perform predictive maintenance, and/or report a real-time inventory. From that, a design and scope can be developed. Questions should be addressed, such as what building functions will be monitored, whether remote access is required, and who will operate it.

- **Components**: Information-gathering IoT devices and sensors vary greatly in capability, technology, accuracy, and communication range, so all of these will need to be considered when selecting these components. The physical infrastructure must be considered as well. Software to drive the digital twin and visualization needs to be determined. Details regarding components and software types are covered in *Chapter 6, The Smart Building Ecosystem*.

- **Capturing the physical environment**: Specialized camera equipment will be required to digitally record each room and system. Each space will need to have images captured from multiple perspectives. These images will then need to be assembled into a 3D digital twin building model. This may seem daunting, but there are special tools and equipment developed to do much of this work automatically.

- **Functionality**: Once the digital twin architecture is completed, functionality and detail must be added. Visual elements will need to be rendered and optimized to provide things such as lighting effects. Interactive nodes will be attached, and the twin will be integrated with the building.

Creating a digital twin is a straightforward process. There are many platforms available to either custom-design a solution to meet individual building needs or off-the-shelf solutions to reduce costs. In the following sections, we will explore a few digital twin use cases to demonstrate the enormous benefits.

Digital twins helping public safety teams

One issue facing first responders today is the lack of accurate and timely information about a building they are about to enter and the emergency they are responding to. Knowing exactly where they and others are located is critical to safety. Smart building digital twin technology is helping them to respond faster, with more information to achieve significant outcomes, including saving lives.

Smart building systems can send a first responder accurate information, such as where the fire is in the building, how big the flames are, and information about chemicals that may be stored close to the fire. Information from motion sensors, heat sensors, video cameras, biochemical sensors, and other smart IoT devices can be delivered directly to first responders while traveling to the building and during their time in the building.

Other benefits and uses of smart building digital twin technologies include the following:

- **Training**: Digital twins provide first responder students with real-world scenarios and place the trainee virtually in a situation they would encounter in the field. This is done either in a training-only mode with guided instructions or in a reality mode, where trainees are hands-on in simulated situations with no guidance.

- **Information**: Building floor plans, mechanical systems, elevators, lighting systems, security systems, and fire systems can be accessed and controlled directly by first responders.

- **Real-time information**: Firefighters, the police, and **Emergency Medical Technicians** (**EMTs**) can accurately assess the emergency conditions of a building and the occupants through real-time sensor data.

- **Communications**: Digital twins support reliable radio coverage for voice and data communications by creating data used to develop simulations. Critical real-time simulations can be conducted with the actual building data during the emergency if communication channels remain open. Ruggedized laptops and handheld communication devices, coupled with video surveillance from dash and body cameras, can feed information to offsite commanders for real-time analysis and direction. Emergency monitoring and early detection devices can help prevent unwanted events.

- **Design**: Digital twin technology used during the design or redesign stage ensures that a building is designed with first responder requirements in mind. Access points, stairways, elevators, water connections, and communications can all be simulated with digital twins. This can assist in knowing where to place sensors.

- **Emergency notification**: Digital twin technology can assist in the design of an emergency notification system, helping to determine where speakers, alarms, strobes, and other devices should be placed. The twin then can synchronize the network of devices during the event.

- **IP-based fire alarm systems**: These IP-based fire systems are networked on the infrastructure and are addressable to provide fire detection at earlier stages, plus the exact fire location. They can be integrated with vital systems, such as sprinklers, video, and access control.

- **Augmented reality**: Augmented reality displays using digital twin technology are being integrated into firefighters' helmets to provide temperature and oxygen levels along with other vital body statics, which can also be monitored remotely.

- **Gunshot detection**: Detection sensors can increase the accuracy to locate active shooters and expedite a response. It also can serve as a deterrent.

- **Post-event review**: Digital twins provide data post-event to conduct root cause analysis and to run what-if scenarios to improve the next response.

- **Avoidance**: Smart grid sensors can drive preventive maintenance that could potentially avoid a disaster entirely. Replacing a part that is about to go faulty may prevent that faulty part from starting a fire.

The ability to provide a secure integrated, IoT-connected building backbone is indispensable. Building end-to-end unified IoT networks that facilitate systems communicating with other systems will let the firefighter adjust the building's air flow rate to control the fire, all while traveling to the building. Seamless interfaces with apps and platforms will require the use of APIs and connectivity.

Smart hospitals using digital twin technology

Subsystems data is combined with real-time interactions between connected devices, people, and processes to deliver the smart hospital digital twin. Intelligent operations, improved situational awareness, real-time information analytics, and orchestration deliver better outcomes for patients and staff.

Today, hospitals are using digital twins to address issues such as the following:

- Long wait times

- Transcription and translation errors

- The downtime of medical facilities and devices

- Interdisciplinary communications and coordination

- Budget and staffing issues to reduce burnout

- Detecting workflow problems before they occur

- Optimizing processes to improve patient outcomes

- Gaining greater clarity into past, present, and future overall performance

Digital twins allow a hospital to analyze the entire picture in real time. They can examine the end-to-end hospital environment and workflows. They can facilitate immediate action if there is an emergency or a deteriorating patient condition that requires a real-time response. The digital twin allows hospitals to analyze the whole picture in real time.

Using digital twin technology at the organizational level, a hospital creates the twin of the hospital building, the administrators, the doctors, and the nurses to gather real-time patient health and workflow insights. Sensors monitor the patients along with the equipment and staff to alert the correct people at the right time for immediate action. Digital twins can also eliminate human interoperability issues, such as automatically triggering an alarm as opposed to waiting for a user to trigger it.

Emergency room wait times can be reduced by removing patient flow and bed management bottlenecks. Code blue emergencies such as cardiopulmonary or respiratory arrest can be predicted and prevented using digital twins. Medical equipment, devices, and facilities can minimize downtime with predictive maintenance practices. Artificial intelligence and machine learning models can be added to detect and predict events in the hospital.

Lowe's stores introducing digital twin technology

Lowe's Companies, Inc. (NYSE: LOW) is a North American home improvement box store and is one of the first retailers to introduce interactive digital twin technology, to combine spatial data with product location and historical order information. Using augmented reality headsets, employees can locate items in the store that may be obscured, optimize restocking strategies by seeing how shelves would look, and suggest changes to store plans. Product arrangement and displays can be viewed and modified before they are implemented in the real store environment. A 3D product catalog is currently under development.

Lowe's digital twin applications include the following:

- **Restocking support**: Lowe's associates can wear an AR headset to view a digital twin hologram overlaid atop the physical store layout. Associates can compare what the shelf should look like versus what it actually looks like.

- **X-ray vision**: Hard-to-reach or obscured items in cardboard boxes can be viewed, instead of having to find a ladder or lift and then move the box to the floor to open and view the product.

- **Collaboration**: Store associates can work with the store's central planners to improve planograms (computer-generated merchandise display plans).

- **Store optimization**: The store can use customer traffic and sales performance data to optimize the customer's online shopping experience. In the store, they can use 3D heatmaps and distance measurements to determine what items may be frequently bought together.

Special attention will need to be given to keeping the consumers' data private through the use of personas or avatars. Retailers can track their products throughout the supply chain to meet consumer demand and eliminate waste.

Digital twin smart building challenges

While digital twin and 3D technology do exist, smart building adoption will continue to face challenges, and many building owners see it as daunting and out of reach. Other challenges include the following:

- **Expensive**: Adopting digital twin use for smart buildings is very expensive. Owners and operators with large portfolios can spread that cost across their portfolio, while smaller companies can choose to pick smaller low-hanging fruit projects and stagger implementations.

- **Lack of standards**: There is a lack of real-time, open, and quality data formats, due in part to the lack of industry standards.

RealEstateCore is a Swedish consortium of building owners, software firms, and research institutions that developed an open source ontology, using Microsoft Azure Digital Twins and its **Digital Twins Definition Language** (DTDL). An ontology is a shared data model (or set of models) and the best practice for a domain such as an IoT system, building structure, and energy grid.

These provide a common basis for modeling smart buildings, and to prevent reinventing the wheel, it uses many existing industry standards such as W3C, BRICK schema, and **Building Topology Ontology** (BOT). Best practices have been developed on how to use the ontology, which are located at `https://github.com/azure/opendigitaltwins-building`.

Summary

This chapter demonstrated how digital twins are the lifeblood of smart buildings by unlocking the benefits of IoT and smart building applications. They create long-term value by solving big problems. They are used to improve efficiencies, detect problems, optimize processes, and innovate for the future. Smart building digital twins will also benefit smart cities and urban planners, allowing them to predict what might happen if a building is constructed in a certain manner.

In the next chapter, we'll begin to explore methodologies for pulling together your smart building project by looking at how to define the project requirements.

Part 3:
Building Your Smart
Building Stack

The complexity of smart buildings can be overwhelming, especially when there are numerous vendors, products, and technologies involved. Smart building stacks can be used to map existing building systems, IoT devices, and technologies for comparison or to identify gaps in a vendor's product or from a customer's requirement perspective.

This part contains the following chapters:

- *Chapter 9, Smart Building IoT Stacks and Requirements*
- *Chapter 10, Understanding Your Building's Existing Smart Level and Systems*
- *Chapter 11, Technology and Applications*

Smart Building IoT Stacks and Requirements

9

The complexity of smart buildings can be overwhelming, especially when there are numerous vendors, products, and technologies involved. Smart building stacks can be used to map building products, IoT devices, and technologies for comparison, or to identify gaps in a vendor's product or from a customer's requirement perspective.

The *smart building stack* has different meanings to different stakeholders, with various types such as development stacks, financial stacks, business stacks, technology stacks, and numerous others. In this chapter, we will introduce a smart building stack developed by several industry leaders called the Smarter Stack. This very versatile tool can help those trying to explain, plan, and implement smarter building initiatives using a single tool. It can be used to provide a perspective for a small smart building project in the built community or to plan and design new and retrofit initiatives.

In this chapter, we're going to do the following:

- After a quick review of the industry's common TCP/IP stack and the concept of stacks in general, we'll introduce a smart building stack
- Provide an overview of Monday Live! and the Smarter Stack industry-developed tool and guide
- Explain how to use the Smarter Stack
- Provide examples of using the stack to identify business categories, connectivity, communications, and cybersecurity requirements
- Review how a high school building can use the stack to determine its requirements
- Show how to use the stack to gain valuable perspectives
- Apply the stack application to financial, technology, standard, and protocol requirements
- Discuss smart building initiatives' different approach styles

What are smart building stacks?

Most hardware and software developers use the term *stack*, such as *XXX stack*, to describe a set of protocols used during their development process. One of the most common stacks in use is the TCP/IP stack, otherwise known as the internet protocol suite, used to send emails and data files. Today's intelligent buildings with building automation systems rely on this TCP/IP stack to communicate between systems.

The TCP/IP stack establishes a set of communication rules and standards between the different technology layers. These rules allow for different products to communicate with each other, as long as they are developed using the same guidelines. There are solution stacks, technology stacks, software stacks, registry stacks, memory stacks, value stacks, and so on.

Our objective in this chapter is to not review all these technology stacks, but rather to introduce a business value smart building stack to use for developing your smart building projects. A fundamental starting point for any smart building initiative, regardless of project size, is to understand what the requirements are. The challenge is that with so many stakeholders and project interdependencies, it becomes very difficult and complex to see a clear path. *The Smarter Stack* provides a visual tool to understand requirements and reduce the complexity of smart building projects.

The Smarter Stack

Monday Live! (`mondaylive.org`) is a group of smart building industry thought leaders working together in an open ecosystem, who exchange ideas, approaches, information, and best practices for smart buildings. Their mission *is as a catalyst to identify, discuss, and clarify relative topics that help drive buildings to be smarter*. This group meets every Monday afternoon to discuss smart building topics freely. They formed during the pandemic to keep moving the industry forward.

They have developed the Smarter Stack and are, in their words, *simplifying the complex landscape of smarter buildings*. It is a framework that highlights all the pieces needed to make buildings smarter, and it is used to design and collaborate on smart building projects. It is an open source planning guide and tool, available to everyone, that visually depicts a building's requirements from several different perspectives. It is used to explain a particular product, technology, and standard as well as to perform gap analysis and more.

With their approval and our great appreciation, this chapter will discuss the Smarter Stack in detail, but first, we would like to recognize the individuals that worked to pull this together during the pandemic: Anno Scholten, Anto Budiardjo, Bill Behn, Gina Elliott, Jim Lee, John Petze, Ken Sinclair, Marc Petock, Roger Woodward, Scott Hoffmann, Steve Fey, and Tracy Markie.

The structure of the building industry is that there are a lot of different players, numerous products, complex technologies, and a variety of solutions and components that address different parts of a smart building. The group thought the best way to discuss the complexity of this industry was to think about it in terms of a stack.

They initially choose the word *smarter*, since the term *smart building* has so many different meanings to different people at different stages, so they wanted the focus to be on making the building smarter and smarter as projects are implemented. The Smarter Stack is a journey and not a destination and was designed to reduce the complexity of smart buildings.

The Smarter Stack begins at a high level by looking at the major components:

- **Users**: These are the people who own, operate, occupy, and gain benefits from the use of the building.

- **Smarts**: This consists of the technology and applications that make the building smarter. There are two parts to the *smarts* stack:

  - **Applications** that make the building smart

  - **Data** that is created regardless of source location

- **Building**: The final component is the building, with its focus on the physical building itself and the systems and controls needed to operate the building.

To make these components work together, the *needs* of each were determined at a very high level:

- **User's needs**: For a user, their need is to be able to go about the building in a safe, healthy, and productive environment.

- **Smart's needs**: On the smart side, apps need to be able to be delivered to users and to access a building's data. Data needs to be gathered, stored, and made available for use by the applications in a secure manner.

- **Building's needs**: A building needs to be available 24 hours a day, 7 days a week.

The Monday Live! team visualized this as shown in the table in *Figure 9.1*:

	Users	All those who occupy, operate, own, or otherwise consume the building.	**Needs** buildings to be operated to be safe, healthy and productive environments.
Smarts	Apps	The applications designed to make buildings smart.	**Needs** to deliver apps to users and access data about the building.
	Data	The data about the building regardless where it is located.	**Need** to securely gather, store and make available for applications and users.
	Building	The building and its control and automation systems.	**Needs** to be available 24/7.

Figure 9.1 – The Smarter Stack's stack of needs

With the fundamental needs identified, what is missing at this point is how these all work with each other, which takes us to the next level of zooming in on the stacks.

At the top of the stack are the users, and there are numerous types of users of a smart building. For simplification, they split these into two groups of users:

- **Purpose users**: The purpose user group consists of owners, managers, and occupants that are interested in the outcomes the building has to offer
- **Operations users**: The operations user group consists of the facility managers, maintenance people, janitorial people, and others that operate and manage a building daily

At the bottom is the building level, and this has been split into two groups as well:

- **Physical group**: The physical group consists of the wood, steel, windows, and concrete that make up the physical structure, along with the necessary equipment, such as boilers, plumbing, wiring, lighting, and HVAC
- **Systems group**: The systems group consists of the control systems and devices required to operate a building and equipment

In the middle is the "smarts" level, which is broken down into four components:

- **Delivery layer**: The first layer is the delivery layer, which is the products, people, and technologies that are responsible for the delivery of the smart to the users (purpose users and operations users). This layer is all about how the smarts are delivered.
- **Apps layer**: The apps layer contains digital applications that are used to improve a building and to make it smarter.
- **Exchange layer**: These are the mechanisms that make the data, apps, and delivery work together. This includes middleware, industry standards, and products that make everything work together.
- **Data layer**: This layer focuses on how data is collected, stored, and normalized, along with the data governance and management processes.

When these are all put together, the eight levels make up the Smarter Stack, as shown in *Figure 9.2.*

Purpose Why is this being done?	The outcomes desired by owners, managers and occupants
Operations How it is operated daily?	The operations to manage the building on a daily basis
Delivery How the smarts is delivered?	The delivery of smarts to the operations and other users
Apps What makes it smart?	The smart digital applications designed to improve the building
Exchange What makes it work together?	The exchange, integration & matching of apps, data, & services
Data What makes it valuable?	The storage, normalization, governance & management of data
Systems What makes it work 24/7?	The automation/control & devices to operate the building
Physical What is being made smarter?	The steel, concrete and equipment that makes the building

(Smarts spans from Delivery through Physical on the left side)

Figure 9.2 – The Smarter Stack

The Monday Live! team compared their Smart Stack to Simon Sinek's Golden Circle from his book *Start with Why*. Sinek purports that to get things done, every organization knows WHAT they do, while some know HOW they do it; however, very few organizations know WHY they are doing it.

The Smarter Stack's purpose level equates to Sinek's WHY to answer the question of why you are trying to make a building smart or smarter. The operations and delivery layers of the stack represent HOW the building is operated daily. The remaining layers (apps, exchange, data, systems, and physical) represent the WHAT. Sinek's theory is that most organizations focus on the WHAT; therefore, the Monday Live! team wanted to make sure the stack paid attention to the WHY and HOW as well.

How to use the Smarter Stack

One of the intentions of the Smarter Stack is to use it as a backdrop to explain how a product or technology fits into an overall smart building. For example, we could place product X in the apps category, which means it is not a delivery smart or an exchange smart. We can clearly delineate that it is an application of some sort. We can continue to use the stack as a backdrop to place technologies and logos.

Another way to use the stack is to take current offerings and place them in the stack to gain insights as to where the building currently stands regarding smarts. Today's vertically integrated **Building Automation System** (**BAS**) offerings are an integrated package that overlays the delivery, apps, exchange, data, and systems stacks as one offering. These systems are proprietary and exchange and talk to their own data, with built-in apps and delivery capabilities all in one box.

The purpose and operations stack's user perspectives have been traditionally determined in-house as determined by the owner's management themselves. With the Smarter Stack, the boxes are broken apart and more standalone solutions are offered. The owner's and occupant's considerations and application requirements determine the purpose stack. Facility management, operations, and maintenance requirements are evaluated today, along with the increased use of digital twin technology, to drive operation stack requirements.

Whereas with the BAS system, manufacturers typically installed and integrated their systems. In the smarter model, many different components and applications are delivered and integrated by a **System Integrator (SI)**, **Master System Integrator (MSI)**, and/or **Managed Service Provider (MSP)**. From a technology perspective, delivery is moving toward a **Single Pane of Glass (SPOG)** and other digital delivery mechanisms. Custom applications are available for single or multiple-use purposes (that is, indoor air quality).

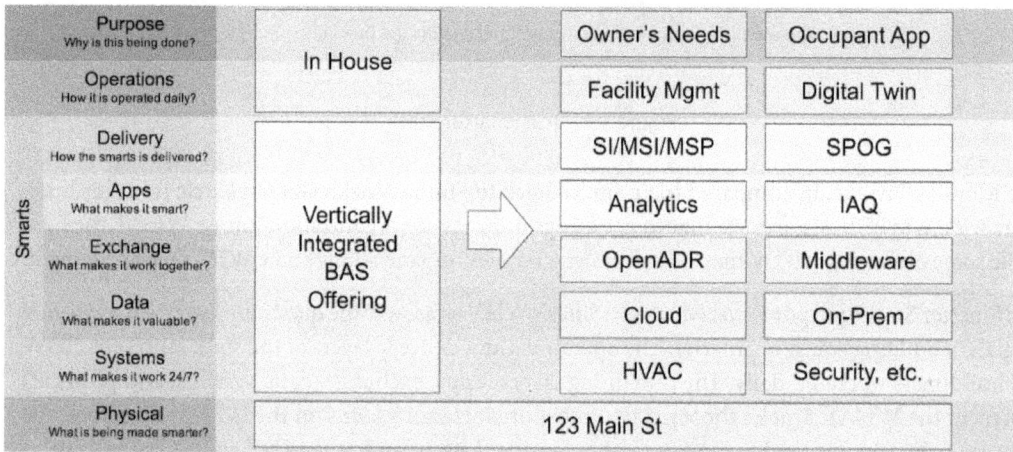

Figure 9.3 – A traditional versus smarter model

Middleware, an open **Automated Demand Response (ADR)** (for energy management), and industry standards are in the exchange layer. In the data layer, on-premise, edge, fog, and cloud computing are all available, and it doesn't matter which is used. At the systems layer, newer, single-purpose, and smarter systems are available outside the box. *Figure 9.3* visualizes this traditional model versus the smarter model.

We are moving from a monolithic, vertical, proprietary, and difficult-to-break-apart model to a model where everything is inheritably composable and broken apart to meet the users' specific needs.

The Smarter Stack use cases

The Smarter Stack was designed as an open source planning tool and guide. By visually depicting a building's requirements, each of the various stakeholders can design and collaborate smart building initiatives using an interoperable approach. Individual focus areas such as business categories, smart building requirements, communications and connectivity, and cybersecurity can each be addressed using the stacks, offering a single easy-to-use platform. In this section, we will look at each of these use-case examples to demonstrate how the Smarter Stack can be used. Your Smarter Stack will mostly likely vary, as these are only examples.

Business categories use case

Figure 9.4 indicates how a smattering of business-related categories can fit into the Smarter Stack. For the users' stack, the **Purpose** layer could have objectives to increase profit in a retail environment, have happier students in a school, or improve wellness and safety for an enhanced tenant experience in a commercial office. At the **Operation** layer, we can include daily facility management, property management, and energy management along with maintenance and janitorial services.

In the smarts stack at the **Delivery** layer, there are a number of ways at looking at who is delivering the service and the various ways it can be delivered. In the **Apps** layer, there are opportunities for energy management, artificial intelligence, and machine learning applications, along with facility management and **Integrated Workplace Management Systems** (**IWMS**) applications. Semantic tagging, middleware, gateways, and integration tools are all part of the **Exchange** layer. Internally, real-time collected data can be integrated with stored and third-party sources at the **Data** layer.

Purpose Why is this being done?	Profit	Happy Students	Tennant Exp	Wellness	Safety
Operations How it is operated daily?	Facility Mgmt	Property Mgmt	Energy Mgmt	Maintenance	Janitorial
Delivery How the smarts is delivered?	Srv Providers	SPOG	MSI & MSP	Consultants	
Apps What makes it smart?	Analytics	Energy Mgmt	AI/ML	FM & IWMS Applications	Vertically Integrated BAS/BMS Providers
Exchange What makes it work together?	Gateways	Middleware	Semantic Tagging	Integration	
Data What makes it valuable?	Cloud	Cybersecurity	Weather	Data Warehouse	
Systems What makes it work 24/7?	Control Systems	Physical Security	IoT/BAS	Lighting Systems	
Physical What is being made smarter?	Schools	Retail	Pharma	Elevators	Airports

Figure 9.4 – Sample business categories

Within the **Systems** stack are all the building's control systems, such as security, HVAC, lighting, BAS, the IoT network, and other systems. The **Physical** layer of the stack focuses on the type of physical building based on the building's function, such as schools, hospitals, retail, airports, and the type of equipment in the building, such as elevators in a commercial building and pharmacy dispensary equipment in a drugstore.

Connectivity and communications use case

The Smarter Stack is not a technology stack; rather, it is a business stack, and therefore, the stack itself does not address connectivity and communications, which we highlighted earlier in the book as critical to smart buildings. The way to address these is to use the Smarter Stack as represented in *Figure 9.5*.

When choosing a certain type of connectivity, for example, we can start again at the top of the stack and ask what the purpose of something is or why something is being done. Are we trying to solve bandwidth, latency, and reliability requirements, or is the purpose just to make it available at an economic cost? At the **Operation** stack layer, we look at how are we going to deliver connectivity and communications capabilities daily through the latest available technologies, and at what costs.

Websites, mobile smart devices, virtual reality, augmented reality, digital signage, and digital twins are a few of the ways to deliver smartness. At the **Apps** layer, consideration is given to what firewalls will be required, the type of **Software-Defined Network** (**SDN**) used, along with types of consoles, **Security Event and Incident Management** (**SEIM**), **Secure Access Service Edge** (**SASE**), and other security frameworks that might be implemented.

Protocol converters, gateways, middleware, and the internet are what make smartness work at the **Exchange** layer. Value is added with different types of data, databases, and servers, along with the APIs, data security, and syslogs (messaging protocols) that are included at the **Data** layer.

	Purpose Why is this being done?	Bandwidth	Latency	Reliability	Economics	Availability
	Operations How it is operated daily?	WiFi	5G	Satelites	QR Codes	Humans
	Delivery How the smarts is delivered?	Websites	Mobile	Digital Twin	VR & AR	Kiosks & Digital Signs
Smarts	**Apps** What makes it smart?	Firewalls	SDN	Management Consoles	Monitoring	SIEM/SASE
	Exchange What makes it work together?	Gateways	Middleware	Brokers	Protocol Converters	Internet
	Data What makes it valuable?	APIs	Data Security	Servers	Syslogs	Databases
	Systems What makes it work 24/7?	Hub/Switch	VPN Devices	NAT	Routers	POTS
	Physical What is being made smarter?	Antenas	Cat 5	Cable (TV)	Twisted Pair	Conduit

Figure 9.5 – Connectivity and communications

At the **Physical** level, we explore what is being made smarter, whether it is a Cat 5/6 Ethernet cable, twisted pair, fiber conduit, antennas, or the devices that make connectivity and communications work, such as **Plain Old Telephone Service** (**POTS**), **Virtual Private Network** (**VPN**) devices, hubs, and switches, along with routers and the **Network Address Translation** (**NAT**) internet standard for **Local Area Networks** (**LANs**).

Cybersecurity use case

Cybersecurity is not a layer in the Smart Stack, as it is more of a vertical requirement that needs to be addressed at every layer. Here again, the stack can be used to describe the cybersecurity requirements, as shown in *Figure 9.6*. Starting with the **Purpose** stack, what is the cybersecurity objective and approach? Is it a defense-in-depth approach, using a combination of advanced security tools and establishing zero trust and security policies, or is it a **Confidentiality, Integrity, and Availability** (**CIA**) triad model to guide policies?

Two-Factor Authorization (**2FA**) and **Multi-Factor Authentication** (**MFA**) are part of the mechanics of making cybersecurity operational, along with the **Security Assertion Markup Language** (**SAML**) and **Lightweight Dictionary Access Protocol** (**LDAP**) authentication methods. Distributed generators using grid supply lines with solar panels (otherwise known as islanding) is one of many options available to supply operational and backup power.

Purpose Why is this being done?	Defense in Depth	CIA/AIC Triad	Zero Trust	Security Policies	Nominal Operation
Operations How it is operated daily?	2FA/MFA	LDAP/SAML	Islanding	Penetration Testing	
Delivery How the smarts is delivered?	SSO	Patches & SW Updates	Scorecards & Security Reports		
Apps What makes it smart?	Next Gen Firewalls	SIEM/SASE	Management Consoles	Threat Intelligence	IAM (Users/Devices)
Exchange What makes it work together?	Firewalls	Data Diode & Air Gap	Security Certificates	Trust Vectors	Asset Mgmt Back/Restore
Data What makes it valuable?	API Keys	Encryption	Syslogs	Backups	
Systems What makes it work 24/7?	VPN/VLAN	TLS Encryption	Secure Config	Obscurity	
Physical What is being made smarter?	Physical Access	Hardware Security Key	Conduit Protection		

Figure 9.6 – A cybersecurity example on the Smarter Stack

Cybersecurity delivery of the smarts engages **Single Sign-On** (**SSO**), scorecards and security reports to understand current strengths and weaknesses, and **software** (**SW**) updates. Like the connectivity and communications example, firewalls, SIEM, SASE, and other security framework considerations will be reviewed for the **Delivery** layer.

At the **Apps** layer, consideration is given to firewalls, the type of SDN used, consoles, SEIM, SASE, and other security frameworks. The data diode and air gap used for network communications data transfer, along with trust factors and security certifications used for authentication, make it all work together at the **Exchange** layer. API keys, encryption requirements, syslogs, and a backup strategy are considerations for the **Data** layer.

At the **Physical** layer, there is conduit protection for wires and fiber physical protection, along with the physical security of a building, its hardware, and its equipment. At the **Systems** layer, which makes everything all work together, we'll determine cybersecurity considerations for the VPN, VLAN, and all other systems, along with a determination of encryption and security configurations.

High school requirements use case

The Smarter Stack may also be used to think about requirements and in this use case, we are looking at a fictitious high school's requirements. In this example, the building engineer is sitting at a table with the school's superintendent, determining why they need to make the school smart.

Once again, starting at the **Purpose** layer to answer the why question, it could be that they need smart scheduling tools to make the building more comfortable for students, or to make the building healthy and safe. From an operations perspective, having school and energy dashboards may be a requirement, along with access to a **Closed-Circuit Television** (**CCTV**) network and an HVAC system, and digitizing trouble tickets.

	Purpose Why is this being done?	Scheduling	Comfort	Social	Critical Safety	
	Operations How it is operated daily?	School Dashboard	HVAC Engineer	CCTV Monitoring	Energy Dashboard	Trouble Tickets
	Delivery How the smarts is delivered?	MSP	Local HVAC	Console		
Smarts	Apps What makes it smart?	DER	WELL	FDD	IAQ	EMS
	Exchange What makes it work together?	Configurator	Integration	DER Agg		
	Data What makes it valuable?	Cloud	Weather	Video Storage		
	Systems What makes it work 24/7?	HVAC	Access	Lighting	Fire System	
	Physical What is being made smarter?	Elevators	123 Main	Sport Stadium		

Figure 9.7 – The Smarter Stack requirements categories

At the **Delivery** layer, the superintendent may like the idea of working with a **managed service provider** (**MSP**) and using their own local HVAC contractor, with everything developed into a dashboard console. Apps that may be identified include **Distributed Energy Resources** (**DERs**), **Indoor Air Quality** (**IAQ**), WELL building certification, **Fault Detection and Diagnostics** (**FDD**), or an **Energy Management System** (**EMS**).

Making the apps work will require exchanges using a configurator, integration tools, and/or DER aggregation, along with data storage and use requirements. At the **Systems** layer, the school has identified what systems they want to bring together, such as the HVAC, access, lighting, and fire control systems, and then they need to determine how these systems will work together. At the **Physical** layer, they required the elevators be upgraded at the 123 Main Street location, and that led to the sports stadium being identified as a candidate to be made smarter.

The Smarter Stack can also be used to look at projects from different people's points of view, from a purely financial perspective and with consideration to technology, standards, and protocols. Each is examined in the following sections.

Using the Smarter Stack to gain perspectives

Depending on who you are or who you might be discussing your smarter project with, you will each have different perspectives about a building and a project, as represented in *Figure 9.8*. The building owner may only care about the purpose and operations of the building and probably leaves everything beyond that to others. The consumer or occupant will care about the purpose of the building and why they are using it, be it for retail, hospitality, commercial office, religious, or medical purposes.

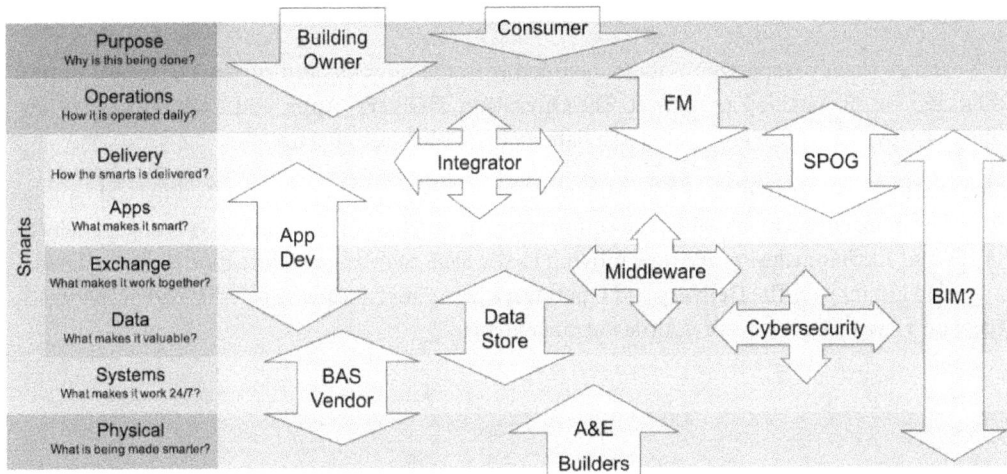

Figure 9.8 – The Smart Stacker perspectives

The **Facility Manager** (**FM**) will undoubtedly focus on how the building operates and functions daily. Integrators will be focused on the delivery of applications to support the operations along with the teams pulling together dashboards on a SPOG.

Applications developers will be involved in the development of the app and a discussion of how it works together. They will also need to be involved in the delivery mechanism and how data is collected and stored. The teams delivering the data storage and management solutions and cybersecurity solutions will not only be involved in the **Data** layer but will also be involved with the systems and exchange.

Middleware (that is, database, mainframe data management, and communications applications) will be highly focused on at the **Exchange** layer while also relevant at the **Apps** and **Data** layers. A BAS providers focus on systems and their integration with a physical building, its equipment, and the data collected. **Architects and Engineers** (**A&E**) and builders will focus their perspectives on the physical building and systems.

The Smarter Stack for financial considerations

Another way to use the Smarter Stack is from a financial perspective. When owners and operators look at a building from a financial viewpoint, they look at **Capital Expenses** (**CapEx**), **Operational Expenses** (**OpEx**), and the revenue or value the building may bring. Most building projects, either new or retrofits, have typically fallen into the CapEx category, since the expense is related to the physical and system items.

Everything else that is required to operate the building falls into the OpEx category. More recently, we are seeing **Software as a Service** (**SaaS**) projects, where the project is to be paid for with OpEx funds. SaaS projects are delivered using cloud-based software and storage on a pay-as-you-go basis.

From a balance sheet perspective, data is included with the physical and systems elements of the building, and are all classified as an asset. The **Operation**, **Delivery**, **Apps**, and **Exchange** layers are classified as expenses, which are treated as a liability on the balance sheet.

Looking at the stack in terms of time engagement is a real eye-opener. Buildings at the **Physical** layer is a 100-year or more engagement, while the equipment and systems are typically 20-year engagements. The **Apps** and **Exchange** layers are now moving to the SaaS monthly revenue model, while data is considered to last forever. The **Delivery** and **Operations** layers are typically self-performed or managed services and viewed as projects or contract arrangements.

	What is it?	Balance Sheet?	Engagement	Procurement	Value
Purpose — Why is this being done?	Add Value		Daily		
Operations — How it is operated daily?			Contract or Project	Bid	$10k-1m/M
Delivery — How the smarts is delivered?		Expense (Liability)		Relationship	$1-100k/M
Apps — What makes it smart?	OpEx		SaaS (Monthly)	App Store	$1-100k/M
Exchange — What makes it work together?				Tech Specs	$1k-$1m
Data — What makes it valuable?			Forever	Capture / Buy	Priceless!
Systems — What makes it work 24/7?	CapEx	Asset	20 Years	RFP or D-B	$10k-10m
Physical — What is being made smarter?			100 Years	Major Deal	$1m-100m

Figure 9.9 – Financials

From a procurement perspective, a building is usually a bided major agreement with the A&E and construction firms. Systems and equipment are awarded through **request for proposal (RFP)** processes, with multiple vendors summiting proposals. Data is collected or captured from IoT devices and systems, or it can be purchased from external sources.

The **Exchange** layer is usually dictated by the technology required to deliver the apps, and the apps are found in the app store from vendors directly. Delivery can be purchased directly or acquired through an existing managed service provider or system integrator. Daily operations for landscaping, janitorial services, and others are typically procured through a bid process.

The stack can also be used to visualize the value for each layer. A physical building can range from $1m to over $100m, while systems and equipment range from $10k to $10m. For the **Data** layer, there is the cost to collect, manage, and store the data, and even a cost to buy external data, but in the end, data is so valuable it is considered priceless. The **Exchange** layer can range from $1k to 100k monthly, while the apps can range from $1 to $100k per month. Operations are bid on and can vary from $10k to $1m per month, obviously dependent on the size of the building and service provided.

The Stack for technologies, standards, and protocols

Another way to use the Smarter Stack is to look at the different technologies, standards, and protocols that may be required for a project and across a smart building. *Figure 9.10* shows an example of some of these. As new technologies and standards evolve, they can be added to the stack to visually represent where they fit in.

Purpose Why is this being done?			LEED	ANSI/APPA 1000-1	WELL	ASHRAE 180	
Operations How it is operated daily?				Safety Standards	NFPA	ASHRAE 55	Digital Twin
Delivery How the smarts is delivered?		HTML	K8S			VR/AR	
Apps What makes it smart?			Linux & Node.js				
Exchange What makes it work together?		Connection Profiles			IP	APIs	
Data What makes it valuable?		DLT / Blockchain	SQL & NoSQL	Haystack			MQTT?
Systems What makes it work 24/7?		BACnet		Modbus		5G	
Physical What is being made smarter?			Building Standards	Zoning	Building Code		

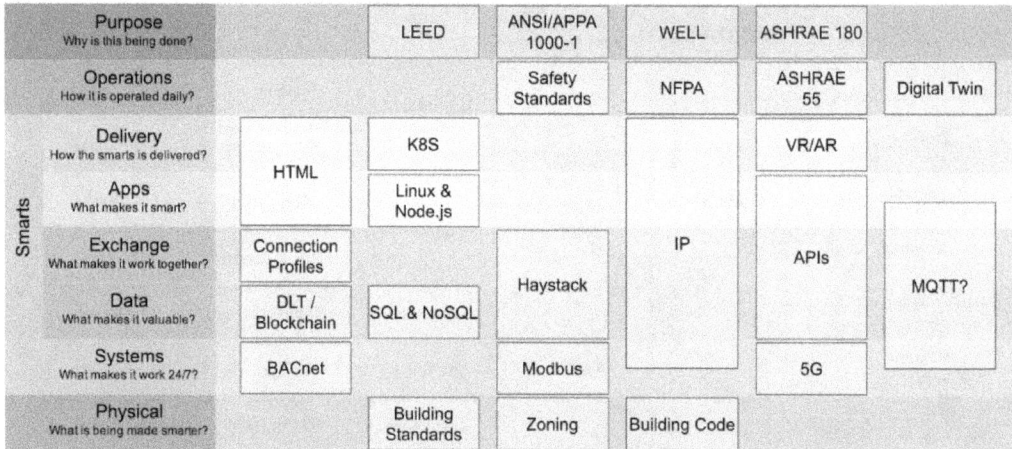

Figure 9.10 – The Smarter Stack to review technologies, standards, and protocols

Many other potential uses for the stack exist, and there really is no limit to how you use it or what perspective you take. Other use examples include looking at the building life cycle, determining the requirements to develop a SPOG, or choosing a vertically integrated BAS.

Individual product and service offerings can be visualized on the stack to show how a SaaS IoT network app not only applies to the **Apps** layer but also the **Delivery**, **Exchange**, and **Data** layers.

What will be the approach, agile or all-at-once?

It is estimated that 80% of the 2050 building stock is already built; therefore, most smart building projects will be used to retrofit existing built assets to improve performance. It's not just about improving the existing systems; it's also about improving an occupant's productivity and quality of experience. The industry cannot wait for conclusive evidence and must push smart buildings faster, harder, and sooner at scale. New design and construction can be built with an eye toward the smart building, but cost considerations usually prevent an all-at-once approach.

To date, building a smart building involved installing individual solutions, such as smart lighting, smart HVAC, and building automation systems. These efforts have been inefficient and disjointed, typically built with silos that are difficult to integrate and scale. Standards are catching up to industry requirements, and technology continues to evolve rapidly.

Understanding the requirements using the Smarter Stack is an excellent way to begin the process of understanding the *what*. In *Chapter 12, A Roadmap to Your Smart Building Will Require Partners*, we will look at recommended approaches to plan and build your smart building initiatives. To reach scale, the industry should adopt **agile methodologies** to achieve the *how*.

The Agile Alliance defines agile as "*the ability to create and respond to change. It is a way of dealing with, and ultimately succeeding in, an uncertain and turbulent environment.*" An approach that started in the software industry is now used across many industries. `Dictionary.com` defines agility (of any kind) as "*the power of moving quickly and easily; nimbleness,*" with a secondary definition of "*the ability to think and draw conclusions quickly; intellectual acuity.*"

There are several agile methodologies such as scrum, **extreme programming** (**XP**), lean software development, kanban, crystal, **feature-driven development** (**FDD**), and the **dynamic systems development method** (**DSDM**). We will not cover these here, other than to mention their existence and to highlight the more popular agile methodology called **scrum**. With scrum, a list of priorities is established by a cross-functional team to deliver increments of solutions in sprints. Problems are identified and corrected early on, and there is strong collaboration between teams.

Using agile methods for smart buildings facilitates more stakeholder involvement to explore what they want from their space. The Smarter Stack helps to identify these stakeholders and their requirements. IoT devices and solutions can collect valuable data to support the use of **virtual reality** (**VR**).

A **virtual reality scrum process** can be used early in a design period to allow users to conceive ideas and then test them in a fully immersive, safe VR space. Users can walk through a completed virtual representation of a smart building, testing it for functionality, identifying issues, and providing improvements and feedback before anything is installed and implemented. This can lower the risk of the procurement process while building a common understanding across stakeholders much faster. Other uses of technology in the design stage will offer additional opportunities to scale more quickly.

Summary

The building industry self-identified the complexity of designing and developing smart building initiatives and, therefore, developed the Smarter Stack guide to visualize and eliminate this complexity. The Smarter Stack is designed to be used across several different perspectives to identify business and financial requirements, along with technology, connectivity, communications, and cybersecurity requirements. Several examples have been provided, but in the end, each smart building project will be as unique as the building itself, and requirements can be reviewed from many perspectives. Finally, agile methodologies should be implemented to scale smart buildings much faster.

Now that we have developed a method for capturing the requirements, the next chapter will focus on identifying and understanding our current building systems and determining what needs to be modified.

10

Understanding Your Building's Existing Smart Level and Systems

In *Chapter 9, Smart Building IoT Stacks and Requirements*, we identified a methodology to define the *why* or purpose for making a building smarter. Since most smart building projects will be in the *built environment* (existing buildings), it's important to understand the current smart state and readiness of the building. In this chapter, we will introduce several industry-smart building assessment programs to determine the current level of a building's *smartness*.

It is also important to understand what systems exist in the building to evaluate whether they will need to be modified, upgraded, or replaced, or if additional technology can be introduced to make that system perform smarter. To continue the journey toward the smart building, an inventory of the current systems, connections, and existing integration will identify *what* the building has as a starting point.

In this chapter, we're going to provide methodologies to do the following:

- Determine how smart the building currently is.

- Perform a smart building assessment using available industry assessment programs. Five programs will be introduced.

- Determine the building's existing systems by identifying eight system categories: civil, structural, mechanical, electrical, energy, plumbing, technology, and communications.

- Suggest a few smart building opportunities for each category.

- Conduct post-assessment activities, such as developing inventory lists and network diagrams.

- Identify system modification, upgrade, replacement, and addition requirements.

How smart is your building

In the previous chapter, we identified a method to identify the smart building requirements from several different perspectives and multiple stakeholders. Recently, several industry associations and alliances across the world have introduced smart building assessment and certification programs. This is an area that I know very well, as I was instrumental in conceptualizing the industry's first smart building assessment and certification program with the **Telecommunications Industry Association (TIA)**.

The idea to develop a certification program initialized from wanting to provide owners and operators a methodology to understand where their building currently stood in terms of smartness, and then for them to develop a strategic plan to evolve that smartness. Many building owners and operators implemented individual projects that added some degree of smartness, such as smart lighting, smart thermostats, and/or energy efficiency improvements, but few understood the impact of a total smart building, since no criteria to measure it against existed.

We began to develop our assessment program to establish a process to benchmark the current state of a building's smartness and then to provide a methodology to measure the impact of future projects. We had to establish a criterion to measure all buildings equally, while keeping it simple enough so that most building owners and operators could perform a self-assessment. Once the assessment criteria were developed, smart building levels were established so that owners and operators could compare their smart building position against other buildings in their portfolio or competitors. Later, certification was introduced beyond the self-assessment to provide competitive market information and differential status.

The following section highlights some of the more popular assessment and certification programs being used today. These are recommended as a starting point to understand how smart your building may already be and to identify what systems exist that may require additions, upgrades, or replacement.

The Smarter Stack

Monday Live!'s Smarter Stack, outlined in the previous chapter, is an excellent tool and methodology to use to understand your building's current state and how the different elements come together. *Figure 10.1* shows the Smarter Stack being used as an assessment tool for an existing fire station building and its smart building project.

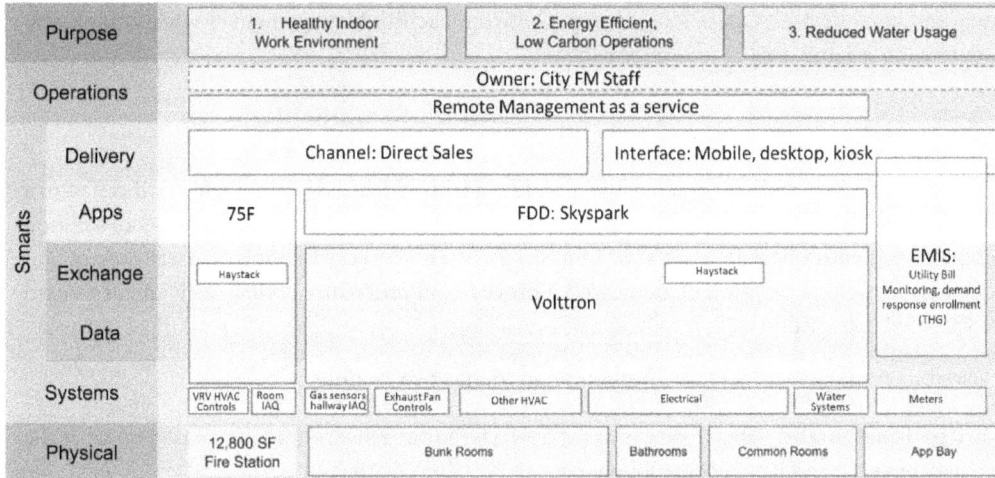

Figure 10.1 – Existing fire station assessment using the Smarter Stack

Starting at the bottom of the stack at the **Physical** layer, we have a 12,800-square-feet fire station that has already been built. Breaking down the **Physical** layer further into areas of the building, we have the bunk rooms, bathrooms, common rooms, and the apparatus bay. At the **Systems** layer, we identified all the systems that were installed, such as the exhaust fans, **indoor air quality (IAQ)**, **variable refrigerant flow (VRF)** control and HVAC systems, electrical, water, and the various building meters.

Moving to the top of the stack, we have the **Purpose** layer; we identified what they wanted in the smart building project. They identified three overriding requirements – 1) a healthy indoor work environment, 2) energy-efficient, low-carbon operations, and 3) to reduce water usage.

The existing HVAC, VRV, and IAQ systems are integrated into a single system by a company called 75F. This system manages the data and exchange processes and contains its own application. Gas sensors, meters, exhaust fan controls, other HVAC functionality, and electrical and lighting systems are all controlled using an application platform called VOLTTRON. This platform is designed for distributed sensing and control exchange along with data storage, which allows developers to work with devices, resources, and each other using an open source, non-proprietary solution. Its drivers are designed for use with Modbus and BACnet protocols.

An **Energy Management Information System (EMIS)** was implemented and is also used to manage utility bill monitoring and demand response enrollment/status. An agile **Feature-Driven Development (FDD)** framework uses an analytics platform, provided by a company called SkySpark, that cuts across the **App** layer to analyze the building's data.

At the **Delivery** layer, it was a direct sale from the company that initiated the design and integration to deliver the project. The interface was delivered via a **Single Pane of Glass (SPOG)**, and mobile, desktop, and kiosk devices. Operations were split between **Facility Management** (**FM**) city employees and remote management-as-a-service providers.

WiredScore

WiredScore is an international company that provides a truly global assessment and certification of a building's digital infrastructure, technological, and connectivity capacity. Their SmartScore solution is designed to identify best-in-class smart buildings. Their scorecard reviews a user's stories and conducts a technology assessment along with a process and procedure review, with an eye toward future-proofed expectations.

The WiredScore assessment and certification program has two sections:

- **User functionality**: This review looks at how a building delivers outcomes based on what the user expects, and it covers the following:

  - Individual and collaborative productivity to create spaces that allow people to be efficient and effective

  - Health and well-being to create safer and healthier work environments

  - Community and service information for the users

  - Sustainability reporting and improvement

  - Maintenance and operations optimization

  - Security for a safe and secure workplace

  Each user story is assessed to the extent that the story has been implemented within the building and to the extent that the story has been implemented to solve issues.

- **Technology foundation**: A review of the technology, processes, and procedures, including the following:

  - Tenant digital connectivity evaluation of the range of connectivity services available to the tenants

  - A building systems evaluation and the integration of software platforms and systems

  - A landlord integration network evaluation of wired and wireless infrastructure installed for the building's systems, and evaluation of its capacity and the range of protocols supported

  - Governance of how the smart building sets its vision and how it defines success measurement and expanded capabilities

  - A cybersecurity review of processes in place and protective steps taken to ensure compliance

  - Data gathering and sharing, system operations, and utilization procedures

WiredScore has four award levels, Certified, Silver, Gold, and Platinum, with the latter being the highest level.

SPIRE Smart Building Assessment™

SPIRE is the industry's first smart building assessment program and was jointly developed by the TIA and the **Underwriters Laboratory (UL)**. The SPIRE Self-Assessment™ online tool evaluates a building using a holistic framework to measure building performance and building intelligence. Six criteria have been defined, and each must work together to deliver a balanced assessment methodology.

The six assessment criteria are as follows:

- **Connectivity**: SPIRE considers connectivity as the fourth building utility and essential to making smart buildings work. Without connectivity, it would be difficult to link equipment, devices, systems, and computing capabilities together for optimization. Connectivity enables applications and systems to transmit, receive, and share data internally and externally. Connectivity is measured by the following:

  - **Media**: Private and public cellular technology, wireless applications, cabling performance, and standards compliance are reviewed. Focus is given to wireless coverage, current and future bandwidth capabilities, low latency, and low-voltage power delivery capabilities.

  - **Coverage**: Sensors and devices should have ubiquitous coverage to support all systems (for example, voice, data, wireless, security, energy, lighting, and building management). Measurement is made by determining the number of wired ports, types, the percentage of cellular and Wi-Fi coverage, E911 support, heat mapping, and by looking at how many carriers provide service to a building.

  - **Security**: Network and infrastructure security are reviewed along with devices, cable entry points, system traffic segregation, people/asset tracking, E911 compliance, and systems backup procedures.

  - **Expansion**: Assuring the capacity and growth potential of power systems, pathways, bandwidth, and wireless coverage.

  - **Resilience**: Ensuring there is a redundancy of power and cabling and de-risking critical events with disaster plans, proper maintenance, and real-time monitoring.

- **Health and well-being**: Post-pandemic requirements coupled with new digital health and well-being IoT devices are driving new smart building healthy space requirements, based on the following factors:

  - **Indoor Air Quality (IAQ)**: The ability to monitor, measure, and manage IAQ using IoT sensors and devices.

- **Thermal management**: Ensuring occupant satisfaction by maintaining comfortable temperature and humidity levels.

- **Visual comfort/light and noise control**: Using IoT sensors and devices to monitor, measure, and manage lighting, acoustics, and vibrations.

- **Water management**: The ability to monitor water quality to ensure there is clean water to manage leaks, usage, and waste.

- **Odor management**: The ability to monitor, analyze, control, and report on offending odors.

- **Life and property safety**: Occupant safety and asset security beyond required requirements and codes are reviewed by looking at the following factors:

  - **Building emergency plan**: Documented evidence of a plan along with accountability levels are reviewed. Integration of systems (for example, the fire system that triggers the HVAC to open or close vents) and the ability of emergency personnel to access these systems when required is critical to the plan.

  - **Integrated system performance**: Integration of life safety systems, risk assessment of the system's performance, and response procedures against established guidelines.

  - **Situational awareness**: Knowing a building's conditions during an emergency – for example, through the use of digital twins – and ensuring proper communication coverage will improve safety and protection.

  - **Emergency communication systems**: A review of adherence to required communications by code and the effectiveness of an in-building system's signal levels, as measured by RF site surveys, along with public safety frequency coverage.

- **Power and energy**: The ability to monitor and manage a building's power and energy use, and electric utility grid response, is measured through the following factors:

  - **Energy use management and analysis**: Assesses the ability to meet energy efficiency and cost goals through the building's capability to track, measure, and manage consumption against predicted models or performance goals.

  - **Demand response and grid interoperability**: Reviews the ability to respond to real-time prices, grid requirements, and incentives from utility providers to reduce operating costs and meet peak load management protocols.

  - **Distributed energy resources**: Measures the ability to distribute energy sources with on-site solar, wind, geothermal, or biomass generators, along with the use of intelligent systems.

- **Cybersecurity**: The ability to manage cybersecurity risk, benchmark capabilities, and set improvement goals in adherence to industry guidelines and best practices. These guidelines focus on the capability to identify, protect, detect, respond, and recover.

- **Sustainability**: This criterion covers many areas, including, water, IAQ, waste tracking, acoustic qualities, and lighting. There are many recognized building sustainability certification programs.

While our focus here is on the assessment of your building, SPIRE offers a Smart Building Verified Mark along with a plaque to display the verified mark. Using anonymized data, comparisons are available by building type and/or assessment type.

Smart Building Alliance

The **Smart Buildings Alliance (SBA)** was established in 2012 in France. The association provides support to all stakeholders of the building industry, including regional players engaged in digital technology, and works with the certifying body Certivéa.

They have developed several frames of reference:

- **Ready to Services (R2S)**: Establishes a baseline of requirements with five principles, broken down into requirements and methods to demonstrate the following:

  - A solution-focused architecture with three independent layers

  - System interoperability

  - An IP standard for the building network

  - Data availability, both inside and outside the building

  - Data security through legal and IT measures

- **Ready to Grids (R2G)**: Reviews the ability and flexibility to interact with an energy distribution system at three levels:

  - **Communicating**: The ability to communicate energy data (consumption, production, and flexibilities) at fixed intervals

  - **Reliable**: Communications reliability and the ability to keep commitments to consumption time, energy, and water production and storage targets

  - **Active**: The reliability and capability of on-demand flexibility for energy production and consumption to react to external requirements

- **BIM4Value Frame of Reference (B4V)**: This is a frame of reference for **Building Information Modeling (BIM)** practices, created to answer questions on value creation, collaboration efforts, and so on, and to establish the many uses of digital modeling and the values expected.

Professional organizations such as CINOV, CNOA, EGF.BTP, FEDENE, FSIF, SBA, and Syntec-Ingénierie have joined forces to write a frame of reference to meet the expectations of the entire building sector. These experts created B4V as a free tool that provides an engineer with support over the entire life cycle of a project, for enhanced control and integration of BIM.

Smart Building Collective assessment and certification

Smart Building Collective is an industry group of partners that perform peer reviews and assessments for in-design projects through occupied buildings. They certify infrastructure and process smart building projects with three independent certified assessors, using a peer review model. These assessors assist through the certification process. Bronze, silver, gold, and platinum levels are assessed and awarded.

This is *not* a checklist process but rather a scoring process that is used to measure the use of technology in the following categories:

- **Building usage**: The focus is on how space, system automation, amenities, and other improvements serve users better. It is about fulfilling the users' needs more effectively while improving performance and cost bases.

- **Building performance**: This looks at a building's overall performance from the systems, amenities, and machine management level to deliver operations optimization and eliminate wasted costs. The focus is on improved efficiencies, consumption, investment optimization, and low environmental impact.

- **Building environment**: The physical environment of a building and materials and the impact on air quality and other health factors.

- **Health, safety, and security**: This category focuses on the use of technology to meet health, safety, and security requirements.

- **User behavior and collaboration**: This explores how a building facilitates user behavior and collaboration to maximize the building's space and purpose.

- **Integrated design and connectivity**: This explores how solutions are integrated with each other and throughout a building. It is reviewed to ensure maximum benefits from the least number of solutions. Connectivity drives continuity and is part of the integrated design.

The Smart Building Collective team along with the three independent assessors also provide three useful peer reviews from the assessment, along with ideas for improvement.

There are many other methods to assess your building, including new technology that performs real-time 3D images and data capture. Consulting firms, system integrators, and system contractors offer individualized assessment services. Regardless of the methodology, framework, technology, or service used, the end goal is to establish a baseline of the building and its systems to work from.

Current building systems

While the assessment methodologies outlined earlier will help you to determine your building's current state of smartness, we recommend performing a complete inventory of all the systems within the building. This can be an exhaustive task, so we have developed the following list of possible systems to identify.

Each of these systems, individually, are candidates for the potential implementation of IoT sensors, controllers, and devices to make them smart or smarter; however, a holistic approach should be used, since a smart building project can benefit multiple systems, and integration of these systems is the ultimate smart building goal. For now, let's identify as many of these systems as we can by breaking them into eight categories. Your building may have more or fewer systems, depending on its size and purpose.

Civil systems

The civil systems for a building typically address the design, construction, and physical elements that comprise it and its surroundings. It deals with the site utilities, the infrastructure to support the building, and landscaping, for example. Civil systems will address the building's orientation, layout, and room configuration, along with features that will make the building attractive, and civil systems usually start with the foundation.

Beams	Framing systems	Roofing system
Concrete	Landscaping	Soils
Curtain walls	Lintels and chajjas	Stairs and lifts
Fencing and external works	Loads and load paths	Utilities
Floor system	Pathways	Wall systems
Foundation	Plinth	

Figure 10.2 – Civil systems

Opportunities exist for the implementation of IoT sensors and devices to monitor and measure building movement and seismic activity. IoT sensors can be added to control soil and landscape water management. Security sensors, cameras, and smart lighting can be added to monitor and manage the pathways. 3D imaging, data collection through sensors, and artificial intelligence can be used to simulate designs before they are implemented.

Structural systems

Columns, beams, trusses, and footers are installed to ensure a building's stability and shape; however, there is much more involved. Consideration might be given to an additional roof structure to accommodate solar panels, satellite dishes, and communication equipment. Pathways to route connections between all the systems and possible building expansion or retrofits are examined here as well.

Building framing types	Footings and foundations
Doors and frames	Interior design
Exterior insulation	Roofing
Exterior wall types	Seismic bracing
Floor systems	Windows

Figure 10.3 – Structural systems

Smart structural systems have a certain level of autonomy, using embedded IoT actuators, sensors, and processors to automatically adjust structural elements, responding to changes in external disturbance and environments. This ensures structural safety and serviceability and can extend the structural service life. Smart IoT sensors can detect damage, and devices using auto-adaptive materials can perform self-repairs. Smart windows can adjust tint, draw shades, and collect and store solar energy.

Mechanical systems

A building's mechanical systems list is usually the longest list of systems; however, the technology list is adding more systems and growing rapidly. These mechanical systems make the building operational and typically suit the functionality of the building. Often, electrical and plumbing systems are included in the mechanical systems list; however, for our purpose here, we have separated those into their own categories. HVAC systems along with ventilation systems are two of the major systems in this category.

Annunciation (alarms)	Elevators	Induction systems
Blowers	Equipment	Laboratory fume hoods
Building central plant systems	Escalators	Life safety systems
Chillers	Evaporative coolers	Mechanical distribution systems
Compressors	Fire protection/smoke detection	Mechanical penthouse
Conduit	HVAC	Motors
Cooling source components	Human thermal comfort	Specialty air systems
Direct expansion systems	Hydronic systems	Variable air volume systems
Dual-duct systems	IAQ systems	Ventilation
Duct system components		

Figure 10.4 – Mechanical systems

A building's mechanical systems present enormous opportunities to introduce IoT smart building solutions. Ventilation and HVAC IoT technology can effectively manage operations to increase comfort while reducing energy costs. IAQ IoT sensors can monitor and manage numerous components of air quality along with temperature and humidity.

Smart duct systems can reduce leakage, eliminate hot and cold spots, and are quieter. IoT fire and smoke sensors can improve fire detection, speed communication, and integrate with other systems for instantaneous responses. **Distributed Antenna Systems** (DASs) can improve critical life safety communications along with IoT occupancy sensors to locate individuals during a building emergency.

Electrical systems

Most often thought of as lights and wall sockets, a building's electrical system consists of many more components. The building's electrical system consists of the means to distribute electricity to every component in the building safely. Wiring, fans, motors, grounding, power distribution systems, and building risers are just a few of the critical components.

AC/DC systems	Electrical wiring	Lighting sources
Conduit systems	Fans	Power distribution
Electric motors	Fire alarm systems	Power system modeling
Electric power	Grounding	Switches
Electric power quality	Lighting applications	Transformers
Electric vehicle (EV) charging		

Figure 10.5 – Electrical systems

Smart lighting projects are often considered the gateway to making buildings smarter. Smart electrical distribution systems add IoT sensors to collect electrical system performance data, smart electrical metering, and software to monitor and perform real-time data analysis, improving efficiency, lowering costs, and meeting regulatory compliance.

Smart whole-building controls monitor, control, and optimize building services such as HVAC, lighting, security, CCTV, audio-visual equipment, and access control. Smart parking and EV charging station equipment with IoT sensors can broadcast to smartphones when spaces are available.

Energy systems

Energy systems encompass aspects of energy efficiency, the use of alternative energy, and energy management systems that are essential to managing and controlling a building's energy needs. Energy consumption, energy conservation, energy recovery, and energy substations are key components of the building's energy system.

Backflow systems	Sewage treatment systems	Vent pipes/ventilation
Drainage systems	Sprinklers	Wastewater systems
Pressure systems	Stack systems	Water cooling/heating systems
Pumps and pipes	Steam and condensation systems	Water heaters and boilers
Re-circulation systems	Storm water systems	Water supply
Septic system	Tanks	Well systems

Figure 10.6 – Energy systems

By accessing actionable data from various smart IoT sensors and meters, buildings can improve efficiency and reduce energy costs. Solar panels, smart windows, and daylight harvesting systems can provide alternative energy sources. Real-time triggers from smart IoT sensors can trigger generators and other backup power systems to automatically and instantaneously kick in when needed.

Smart energy management systems have become one of the biggest smart building deliverables for operational efficiency. IoT devices provide valuable data, used for predictive maintenance to reduce downtime, enable the automatic discovery of potential issues, and increase productivity.

Plumbing systems

Building plumbing systems are a lot more than sinks, toilets, and everything in a restroom. Hot water heating systems, roof drainage storm piping systems, water softening systems, and fire protection sprinkler systems are all included. Complex piping systems, water conservation, water pumps, and re-circulation systems may be required as well.

Backflow systems	Sewage treatment systems	Vent pipes/ventilation
Drainage systems	Sprinklers	Wastewater systems
Pressure systems	Stack systems	Water cooling/heating systems
Pumps and pipes	Steam and condensation systems	Water heaters and boilers
Re-circulation systems	Storm water systems	Water supplies
Septic system	Tanks	Well systems

Figure 10.7 – Plumbing systems

Building flooding is one of the top insurance claims each year, and it is not related to natural disasters. Typically, building flood claims result from broken or burst systems. IoT sensors can play a big part in reducing these issues by sensing leaks and responding accordingly by shutting down the water supply. IoT pressure sensors can monitor and report conditions in real time and prevent incidents through predictive maintenance programs. Water quality IoT sensors can ensure that the water is clean and safe. IoT sensors on vent pipes and ventilation systems can ensure there is no blockage or issues. Smart restroom IoT sensors can notify when restrooms need to be cleaned and stocked.

Technology systems

Technology building systems include everything **Information Technology (IT)** and **Operational Technology (OT)**-related to managing and operating a building. In the past, technology systems were considered a subset of electrical systems, but with the explosion in IoT, new smart technologies, and smart building applications, it makes sense to separate them. The technology systems category covers a range of components, such as access systems, cameras, security, IT infrastructure, and **audio-visual (AV)** equipment.

Asset tracking systems	Centralized clock systems	Robotics
Audio visual equipment	Computing systems	Security systems
Building automation systems	Data networks	Servers
Building management systems	Digital signage	Software
Building purpose equipment	Exchange systems	Space planning systems
Cameras	Energy management systems	Television systems
Card access systems	Enterprise systems	Wayfinding/navigation systems
CCTV	Information technology systems	Workflow systems
	Operational technology systems	

Figure 10.8 – Technology systems

Obviously, technology systems are prime candidates and centrally involved with the implementation of IoT sensors, devices, and smart building technologies. Asset tracking, building automation, building management, security, wayfinding and navigation systems, and others require IoT sensors, connection, and computing to perform their smart building functions.

IT and OT systems will need to converge to support the networking, communication, and computing requirements for smart buildings. Workflow systems are becoming smarter, with sensors distributed throughout a building to record activities and completed work. New IoT technology solutions continue to be introduced with recent additions, such as occupancy sensors, IAQ sensors, smart restroom sensors, and gunshot detection sensors.

Communication systems

Communication systems were recently included in the technology systems category, but again, we have decided that it needs its own category, based on recent technological advances and its rapid growth. What started as hardwired telephone switchboards and telephone stations continues to evolve with more sophisticated wireless systems. Copper wiring has evolved from single, twisted pair, and Ethernet cables to optical fiber and wireless systems.

Cellular coverage has moved from outside the building to indoor systems such as DASs, **Bi-Directional Amplifiers** (**BDAs**), small cells, signal boosters, and neutral host systems. Wireless systems have evolved beyond cellular to include a lengthy list of technologies, such as Bluetooth, LoRa, Zigbee, CBRS, Z-Wave, and EnOcean.

Bi-directional amplifiers	EnOcean	Private networks
Bluetooth	GPON optical fiber networking	Signal boosters
Broadcast systems	Intercom systems	Small cells
Cellular coverage system	Internet, gateways, and routers	Telephone system
Communication closets	IoT network	Wireless internet system
Distributed antenna systems	LoRa/LoRaWAN	Zigbee
Emergency communication systems	Neutral host systems	Z-Wave

Figure 10.9 – Communication systems

Communication systems are instrumental to IoT data collection, connectivity, computing, and systems response. Low-power IoT networks will connect these devices and systems. Buildings will have multiple communications technology options, and most will use a combination of different technologies. Legacy and emerging technologies will connect and interact with each other.

Smart building connection, automation, and control rely on communications networks, as do mission-critical building functions. Integrated copper and fiber cabling systems combined with wireless IoT access points enable smart building functionalities, such as energy management, lighting, access management, security, and maintenance. **Machine learning** (**ML**) and **artificial intelligence** (**AI**) building insights will be collected through these communication systems.

A post-smart building assessment

Once a building assessment is completed and all the systems identified, a comprehensive list of these systems and assets should be developed, capturing the following information:

- The location, system vendor name and model number, age, condition, any software along with version/release information, and the expected useful life should be recorded.

- Note any systems that are reaching (or exceeding) their expected operational life

- Indicate any condition-related or operational-related issues

- Identify compliance-related issues that may need to be corrected

- List any functional concerns related to the systems and equipment, and identify any repairs that may be needed

- Estimate the cost to fix issues or upgrade equipment

- Prioritize any problems identified according to the severity level

- Identify any remedial actions to improve functionality, including renovation or modernization upgrades

- Collect and list information regarding any projects already recommended to fix issues and upgrade equipment

- Note any other unique requirements of your building

With the list of systems and their current condition completed, the next step is to understand how these systems are currently connected by developing a network diagram and determining their capability to be modified, upgraded, or replaced to meet your smart building goals.

Network diagram

It is crucial to understand how a building's network(s) are connected and configured. Therefore, it is important to inventory the building's structured wiring, cabling, fiber, and power systems, along with telephone and communication closets. Detailed network topology and network diagrams should be developed to visualize the current state of the building's network(s) and to create a long-term plan.

This visual diagram of the network will outline devices, pathways, and connections, and it will indicate how data flows within the network. There are two types of network diagrams:

- **Logical network diagrams**: These diagrams will display the inner behaviors of a network and will include routing protocols and subnets. This is the most common method used, and it reveals how information flows through the network.

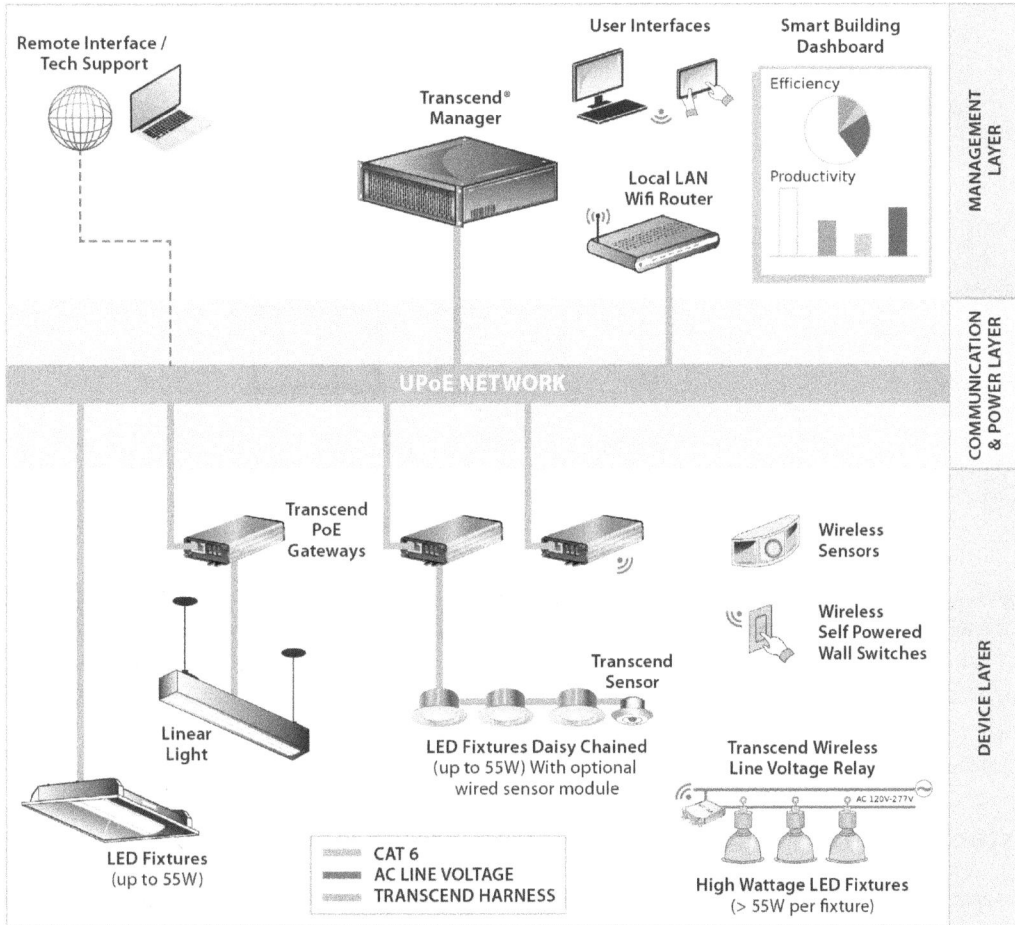

Figure 10.10 – A logical network diagram example (source: Molex Transcend® Network Connected Lighting)

- **Physical network diagrams**: These diagrams show the physical components of a network, such as the cables and switches, and indicate how they are arranged.

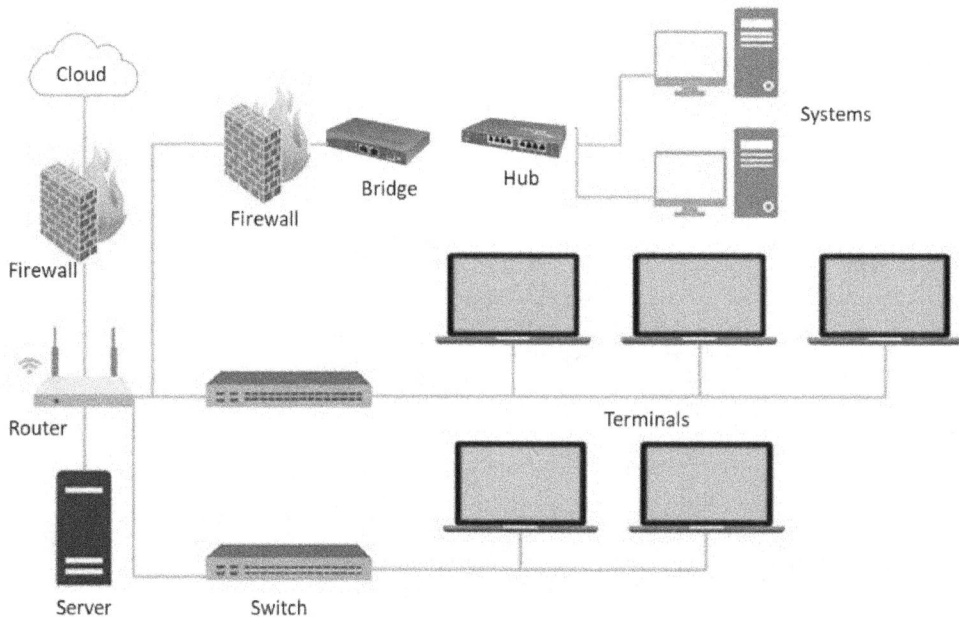

Figure 10.11 – A physical diagram example

To gain a complete picture of a network, both a logical and physical network diagram should be completed. There are different methods to create network diagrams – manually, automatically, and semi-automatically. Numerous off-the-shelf software solutions are available, such as industry leaders SmartDraw, SolarWinds NTM, Intermapper, Creately, Lucidchart, LanFlow, and Microsoft Visio. Whichever is chosen, the software should allow flexibility within the network to adjust with expansion and changes to the network, constantly keeping the diagram updated.

System modification, upgrade, replacement, and addition

In *Chapter 9, Smart Building IoT Stacks and Requirements*, we defined a methodology for developing the requirements to make the building smarter. From that, a comprehensive list of desires and requirements should exist. The next challenge will be to determine whether these smart requirements can be delivered through existing systems or whether a modification, upgrade, replacement, or addition is required.

In many cases, it may be as simple as adding an IoT sensor, actuator, or another device to the physical item or system that is being made smarter. For example, IoT sensors can be added and connected to the existing HVAC system to measure and control humidity and CO_2 levels.

In other cases, the existing system may require modification, such as changing out the filter container compartment with a new container compartment that has built-in sensors to notify when the filter requires replacement. Many existing vendors wanting to extend the life of their systems are exploring ways to offer upgrade capabilities. Dated systems may simply have to be replaced.

The list of smart requirements should be reviewed, and each existing system that may be impacted should be evaluated on its capability to be modified, upgraded, or replaced. Obviously, costs versus return will play a critical role in the final determinations.

Summary

It is important to understand the current state of a building before undertaking any smart building project. Several recently released smart building assessment methodologies and programs have been introduced to help determine the building's current level of *smartness*. It is also critical to understand what systems, connections, and integrations may already exist. This chapter provided an eight-category guide to identify, inventory, and record all the possible systems that exist in the building.

Now that we have developed methods for gathering smart building requirements, performing an assessment of how smart a building is, and inventorying current systems, the next chapter will focus on identifying and understanding the technology, services, and applications that will be required to make the building smart.

11

Technology and Applications

In previous chapters, we examined what a smart building is, how IoT contributes to making buildings smarter, many of the technologies and protocols that are used, and how to define smart building requirements, and we identified the different building systems that could benefit from smart building solutions. In this chapter, we will begin by summarizing the typical hardware requirements to make a building smart.

We'll then examine an extensive list of smart building application opportunities that are available to deliver the smart requirements, along with examples for each. These should be compared to your requirements list. We'll define and explain the role of middleware in delivering these applications. The chapter will conclude with a review of the codes, standards, and guidelines to be considered to prevent becoming locked into proprietary solutions that may prevent expansion later.

In this chapter, we're going to do the following:

- Summarize the hardware required to deliver IoT network smart building solutions
- Provide a list of common smart building applications that can be used to match your list of requirements
- Define middleware and its role in delivering smart applications
- Examine industry codes, standards, and guidelines recommended for best practices

Hardware technology

Smart buildings are becoming increasingly smarter through the addition of hardware technology and applications far beyond **building automation systems** (**BASs**). Smart buildings use advanced technology to collect and communicate information about the building and the system's performance. To do so, certain hardware components are required to create a network to collect data, connect systems, perform analytics, and deliver appropriate responses. The typical smart building network will contain the following hardware components:

- **IoT sensors**: A comprehensive network of IoT sensors and devices that monitor and register events from the environment and collect real-time data from physical assets, systems, and equipment, such as temperature, light, air quality, motion, and more.

- **Connectivity**: Connectivity is the key to the network and allows for the exchange of information across devices and systems in real time. There are many ways to deliver connectivity through wireline, fiber, and wireless technologies.

- **Computing**: Processing or computing either locally, on the edge, or in the cloud performs analytics and delivers commands back to the physical device to respond to commands.

- **Output devices**: Actuators, relays, and similar devices that react to instructions issued by the controllers to cause the physical item to deliver an appropriate response or action:

  - **Actuators**: Actuators receive instructions from other IoT sensors and devices. They have the capability to accept instructions and make real-time physical changes. Actuators act in the opposite way to sensors; where a sensor detects, an actuator will act. Valves, motors, and electric switches are actuators.

  - **Relays**: Relays are electronically operated switches with a set of operating contact terminals. They are typically used when the application needs to switch from low to high voltage (or vice versa) along the same circuit. In a building application such as HVAC, the power temperature sensors require high levels of amperage that exceed the capacity of the wiring. Relays provide amplification to convert the current to a higher level.

- **User interface**: To wrap this all together, a visual mechanism provides dashboards and controls on displays, computer screens, and ultimately, on a **single pane of glass** (**SPOG**), usually a tablet or smartphone.

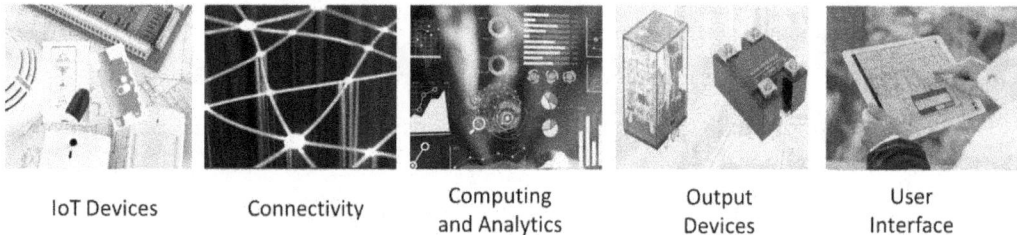

| IoT Devices | Connectivity | Computing and Analytics | Output Devices | User Interface |

Figure 11.1 – Smart building hardware

In addition to the main hardware components of the smart building network, there exist other hardware technologies that may be integrated as needed:

- **Smart metering**: Smart meters track resources that are being distributed and consumed. Precise energy accounting and demand forecasting can help streamline operations. I/O voltage and current readings, water flow, pressure, and consumption trends are provided.

- **Devices**: IoT devices are physical hardware mechanisms that collect and communicate data over a network. They are typically programmed for the application they are serving, such as the following:

- **Point-of-sale devices**: Merchants, restaurants, and other retailers have IoT devices for their **point-of-sale (POS)** terminals. Typically, they are used for their payment terminals to process payments immediately and to collect real-time data about the sale and the customer. They are also used for inventory management and automatic re-ordering. These can be fixed devices at the register or hand-held devices, such as the devices customers can use to pay for their meals at their tables.

- **Medical devices**: The healthcare industry uses connected heart monitors, blood sugar sensors, and internal stomach sensors (to monitor the patient's stomach contents). These devices allow doctors to monitor remotely away from the medical setting and provide more accurate data.

- **Wearable devices**: Smartwatches, earbuds, fitness trackers, and smart glasses encourage wearers to exercise regularly.

- **Smart home devices**: TechJury reports that American households have an average of 22 IoT devices. These include intelligent lighting, smart speakers, smart door locks, and connected smoke detectors. Smart IoT appliances include smart thermostats, environmental devices, and smart refrigerators.

- **Security devices**: These devices allow businesses and homeowners to monitor their buildings remotely.

- **Virtual reality (VR)/artificial intelligence (AI)**: VR and AI are often associated with gaming applications; however, these can be used for more practical purposes in smart buildings. Designers use AR to model rooms and to choose between remolding and reconstruction options. AR is used to visualize what a space may look like when constructed and before the money is spent.

- **Management and automation systems**: **Building Information Modeling (BIM)** and BASs unite machines and humans to automate building management processes.

- **Drones**: Drones are used to make panoramic shots and videos and assist smart building management in areas it may be difficult to reach. They are used for monitoring construction progress, security, product availability, and other activities that are not freely accessible to humans.

- **Robots**: The **Internet of Robotic Things (IoRT)** brings together robots and IoT to monitor events, merge sensor data, and use distributed or local intelligence to decide a course of action to control the physical world. Actuators and sensors typically handle specific tasks, whereby robots can react to unexpected conditions.

IoT hardware comes in many forms to sense and collect information from the physical world for processing and analyzing the data. This hardware is at the very heart of the IoT network, and choosing the right hardware that can be easily integrated with other hardware and software will ensure your smart building network evolves and scales.

Smart building applications

"There's an app for that" is now becoming a common response to requests for smart building technology solutions. With IoT sensors, computing, connectivity, output devices, and display technology hardware components identified already, the software will now be required to drive the applications to make the building smarter.

While we have highlighted numerous smart applications throughout this book already, what follows is a comprehensive alphabetical list of applications that should be considered when reviewing your smart building requirements. There are software programs available throughout the marketplace that can support individual or multiple applications, and new applications continue to be introduced. Some apps are native to the system it supports, while others are delivered as **Software as a Service (SaaS)** over-the-top solutions. Match your requirements to this list of smart building applications or gain new ideas for making your building smarter:

- **Access management**: Occupancy sensors, facial recognition sensors, cameras, card access readers, smartphone QR codes, smart badges, or other IoT device solutions coupled with a software app approve building entrance requests and open the door when criteria are met.

- **Activity-based workplace (ABW)**: Open spaces, hot-desk layouts, and activity-based work settings use IoT sensors to monitor and communicate space status, cleaning requirements, and maintenance needs, all driven by an app.

- **Amenity management**: IoT sensors and a tenant smart app provide real-time access to a long list of building amenities. Wellness center, yoga studio, and workout facility occupancy information can assist tenants in checking occupancy and scheduling their workouts. Information is provided regarding public transportation, ride-share services, taxis, and limo services for the building, along with pick-up and drop-off location information with IoT cameras for real-time status checks and safety purposes. Tenant lounge, food, beverage, dining, entertainment, transportation, retail, and other building options can deliver real-time occupancy and scheduling. Indoor air quality sensors can provide real-time air quality scores and adjust air ventilation as needed.

- **Asset management**: IoT sensors are placed in or on each asset to provide real-time positioning, location, and tracking, and this data is delivered to a software app for analysis, storage, and reporting. They collect information associated with each asset and assign them by department or property to manage, track, evaluate, and assign costs appropriately. Vendors are assigned to the assets to capture support detailed information for easy and efficient vendor support. Cost information is assigned for cost tracking and allocation and to assist in making information-based decisions regarding the building's assets.

- **Automated controls**: Controls allow an operator to access, monitor, and manage the building's connected systems for such things as lighting, HVAC, electrical, security, and access systems by using data from occupant behavior, environmental conditions, and more.

- **Bike storage**: Smart bike storage offers secure innovative bike parking to promote cycling and free up vehicle parking spaces. Secure access, cameras, IoT sensors, and a smart app control who has access and provide real-time space availability.

- **Building management: Building Management Systems (BMSs)** use IoT sensors and controllers to monitor and control all the building's various systems. These systems typically include HVAC, safety and security, cameras/CCTV, fire/alarm, lighting, water, gas, other fluids, networking, communication, and others. These systems provide a wealth of information about how the building is performing and use software apps to process the data and make changes for efficiency improvements and cost reductions.

- **Building modeling**: The BIM app supports the creation of a digital twin record of the facility information to store digitally. These include blueprints, emergency plans, plumbing and electrical installation drawings, 3D laser scans, and building thermal imaging surveys. This saves time, reduces rework, restores missing data and drawings, reduces liability, and minimizes risk.

  Any type of building can be modeled and scanned; a **computer-aided drawing (CAD)** rendering is produced. 360-degree panoramic images are also available. These can be used to reverse engineer an as-built building. Detailed mapping is achieved using 3D scanning techniques, drones, **unmanned aerial vehicles (UAVs)**, terrestrial scanners, and digital photogrammetry.

- **Communications and networking**: Smart buildings use IoT technology coupled with the latest communication systems and applications to deliver a highly secure system that provides coverage to 99.9% of the building, including elevators, stairwells, basements, and the parking garage. DAS, small cell, Bluetooth, Wi-Fi, fiber back-haul, and other technologies deliver high bandwidth while reducing latency for overall control.

- **Conference rooms**: Smart conference room apps and IoT devices help to keep the meeting productive. Users can schedule the room using a tenant application provided by the building manager. Lighting levels and temperature can automatically be set to the user's profile preferences and adjusted via a tenant app.

  With collaborative conferencing technology and AI tools, presentations are automatically loaded, audio and visual systems are turned on, and the conference call autodial is launched. Wireless broadband and airplay allow all to share data in real time during the meeting.

Figure 11.2 – Smart conference room

The smart jamboard captures all meeting information and notes, and the auto transcription system automatically captures and develops meeting notes that are sent to everyone immediately at the meeting's conclusion.

- **Contact tracing**: Many different contract tracing apps are used in response to the pandemic. Typically, they use GPS or Bluetooth IoT technology along with algorithms to detect potential exposure and notify users. They can also anonymously inform the larger community of any potential exposures.

- **Cost optimization**: Building usage patterns analysis is used to optimize HVAC operations, adjust the building's maintenance practices, match occupancy patterns to use projections, enhance space utilization efficiency, and more. Some cost optimization apps are standalone apps, while many others are embedded into other available smart building apps. All apps require access to data collected by IoT sensors and devices.

- **Custom experiences**: Using real-time building data and APIs, custom applications can be developed to deliver one-of-a-kind user experiences. Carnival Cruise's xIoT focuses on enhanced guest-crew interactions to perfect the guest experience. Tenant applications are typically customized with building-specific options and unique offerings and are often branded by the building operator.

- **Delivery**: IoT sensors, coupled with a smart delivery app, assist drones and robots to deliver items throughout the building. Supply chain systems use these sensors, devices, and apps to inventory, track, locate, and deliver parts and equipment in manufacturing applications.

- **Design**: Smart building technology design apps are used in the design of new buildings and for the retrofit of existing buildings. IoT sensors and devices collect building information used to create 3D images and digital twins for redesign purposes. IoT-data-fed AI apps create building simulations to visualize what the building or space might look like before starting construction.

- **Document library**: A smart building library app collects important documents, contracts, leases, forms, maintenance manuals, and other building-related documents. These documents can be updated in real time with data collected from IoT sensors.

- **Door/cabinet open/closed detection**: Intrusion IoT sensors are used to detect the open/closed status of doors, windows, drawers, cabinets, and other open/closed sensing requirements. Activity can be reported, and alerts sent to remote smart devices, such as smartphones via an app. These are also used for security, tracking, and to proactivity prevent damage, spoilage, or theft.

- **Electricity and power**: These apps monitor and manage electric usage in real time to ensure availability, even distribution, and immediate responses to interruptions. Load balancing can be used to streamline energy use and cut costs. Real-time reports and notifications of power fluctuations, outages, or surges are provided. They can send and receive commands to enable or disable electronic devices easily and manage peak power in real time. Trending data supplied by IoT sensors and devices is used to optimize electric usage and to manage the smart electrical distribution platform.

- **Electric vehicle (EV) charging**: The EV app indicates EV charging station locations, availability, scheduling, and current charge level information. They perform smart charging with charging options and load management options. They can arrange charging during off-peak times and adjust charging levels to contribute to grid balancing. Smart charging could avoid costly upgrades to the building's electrical distribution system by managing charging patterns.

- **Energy management**: The **Building Energy Management System (BEMS)** is an app to manage, monitor, and control energy to reduce energy consumption. The BEMS starts with IoT sensors everywhere to measure temperature, changes in voltage, vibration, spikes in energy usage, humidity, motion, light, smoke, acceleration, chemicals, pressure, and other changes that can be monitored. Sensors may be placed throughout the building, on machinery, in lighting systems, in HVAC and ventilation systems, on hot water heaters and pumps, in refrigeration units, and more. The sensor is either connected to a device or is part of the device. Energy use analytics are used to support installing energy-efficient systems and smart meters.

- **Elevators**: Sensors detect a known individual upon entering a commercial or residential building or in the general smart elevator area. The occupant may also use a smart elevator app on their phone to summons the car. Using the occupant's stored information, the smart elevator and the app both flash a personalized welcome screen, select an elevator car, open the door automatically, and deliver the occupant to the correct floor, all touch-free for added safety.

Digital information is provided in the elevator or via the smart elevator app highlighting that car's indoor air quality score, the day's building events, and announcements, along with the cafeteria lunch special. Elevators can be remotely controlled or monitored by Wi-Fi, LTE, or

Bluetooth connectivity from a smartphone application. Smart elevators are more energy-efficient by reducing the number of stops. Predictive maintenance alerts notify operators about required maintenance, potential issues, and other important information.

- **Environmental impact reduction**: Using IoT sensors and smart applications, indoor and outdoor environmental conditions are analyzed, along with occupant behavior and other data to optimize energy and utility consumption to reduce emissions and carbon footprint. This app will usually interact with the energy management app.

- **Fire safety**: Fire safety apps coupled with IoT smoke and fire detection sensors trigger alerts, notify first responders, and trigger responses such as activating the sprinklers. These sensors include addressable pull stations and call points. Addressable modules deliver accessibility to conventional devices such as flow switches. Photoelectric sensors measure the number of airborne particles and their type to sense the presence of fire. Heat sensors detect fire by using the rise in temperature, and sound sensors use low-frequency signaling to monitor unattended areas.

- **First responders**: IoT sensors, actuators, controllers, smartphones, and other devices, coupled with in-building communication systems and notification apps, provide real-time data to first responders.

- **Gamification**: **IoT-enabled Gamification** (**IeG**) is the convergence of gamification and IoT through a smart gaming app. Using IoT sensors, beacons, and tags on the employee's ID badge, points can be awarded for recycling, fitness activities, use of public transportation, or even participation in lunch-n-learn sessions. IoT can track energy-related activities and provide rewards for taking actions to use less energy, such as turning off lights and better utilization of temperature. Smartwatch devices and apps that measure vital signs can be synchronized with sensors in the building's fitness center to provide progress and summary reports.

- **Guest/occupant registration**: Guests may pre-register with a registration app, or they may be directed to a reception desk, a kiosk, or a phone system to complete on-site registration. For existing occupants entering the building, a wealth of stakeholder-approved information is accessed from their IoT-enabled access credentials.

- **Health and wellbeing**: IoT sensors and smart apps are used to measure and promote safe physical distancing. Smart lighting, HVAC, indoor air quality, and CO_2 measurement, all contribute to wellbeing. Space optimization and space cleaning notification solutions provide peace of mind for occupants. Sensors can trigger UV lighting and ionization systems to turn on when space is unoccupied and provide data reports via a smart app.

- **Hospitals**: IoT devices and various apps optimize the tracking, usage, and maintenance of medical equipment. They can facilitate the registration process and manage comfort settings. Wayfinding and navigation help visitors locate rooms and amenities. Patient monitoring, medicine management, and intelligent emergency systems can all benefit from IoT solutions.

- **HVAC**: IoT solutions monitor the system's performance and detect issues before they become major problems. Real-time data using IoT sensors can deliver a more condition-based method for preventative maintenance. IoT sensors installed on HVAC systems and connected to a smart HVAC app can monitor usage trends and, when coupled with weather predictions, can improve energy efficiency.

Remote system monitoring and management can be performed using a smartphone app or website portal. Leak detection, vibration, fault tolerance, and pressure sensors can provide immediate notification of issues. HVAC filters with built-in IoT sensors can notify maintenance teams when they need to be replaced. Air and surface disinfectants are designed to mitigate germs and are monitored with IoT sensors. Most HVAC businesses are required to inspect equipment periodically and document compliance. With IoT sensors, real-time data is recorded via the smart app, which can produce and deliver reports automatically.

- **Indoor Air Quality (IAQ)**: IoT sensors are located throughout the building for air quality monitoring and management. Air purification systems use sensors to monitor and trigger the amount of outside air entering the building. Carbon dioxide and carbon monoxide detectors, temperature, humidity, particulate matter, total volatile organic compounds, and environmental elements are measured and reported using IoT sensors and smart building apps.

- **Internet access**: Many building operators will need to provide building-wide internet access. While copper and fiber have been the traditional method, IoT cellular and other wireless solutions such as Bluetooth, LTE, and 5G can deliver high-speed Wi-Fi access. Apps deliver access and usage data.

- **IT/OT convergence**: Any building that uses IoT devices to collect and transmit data to monitor and control its systems and equipment will experience IT/OT convergence. This occurs when **information technology (IT)** systems are integrated with **operational technology (OT)** hardware systems to collect, connect, and analyze data. The ultimate smart building goal is to bring all the building's data together on a single device using an integration and reporting app.

- **Janitorial and cleaning**: IoT devices are used to notify crews of cleaning services required or recently completed; for example, a work area and computer have been cleaned and the scheduling software app has been updated to allow the next person to use the space. Housekeeping, groundskeeping, landscaping, janitorial, and custodial services are included as general maintenance activities.

- **Landscaping**: IoT sensors monitor the grass, plants, and soil to adjust irrigation and fertilization requirements and provide dashboard information via an app.

- **Leak detection**: Water detection sensors, acoustic sensors, and flow meters are used to detect water leaks. IoT sensors can connect to almost any system and measure airflow leaks in air ducts, and gas, chemical, and fluid leaks. Real-time alerts, preventive maintenance, and analytics apps can provide additional information such as flow direction and rate, temperature, humidity, and other valuable data.

- **Life safety**: Traditionally, life safety systems were isolated proprietary systems within the building. With IT/OT convergence, IoT devices, and smart building apps, these systems are being consolidated into a single application. These systems include fire alarm strobes, heat detectors, fire suppression systems, water flow sensors, and more.

- **Lighting systems:** Occupancy sensors are used to turn lights on and off when presence is detected. Tenant apps allow occupants to adjust lighting preferences. A daylight harvesting system can reduce energy consumption by using daylight to offset the amount of electric lighting needed by dimming or switching the amount of light based on the actual daylight available. LED bulbs, voice activation, and motion sensors, along with lighting system management software apps, allow for additional scheduling and control capabilities.

- **Locker storage:** Smart locker systems and automated package storage systems streamline mailroom operations, improve security, and create an audit trail, while providing immediate notification to the occupant via an app. Smart locker systems reduce the risks of theft, misplaced, or stolen packages or other assets. They reduce mailroom overcrowding by maintaining occupancy sensing and safe distancing requirements. Office buildings, education campuses, and large complex centers can streamline their mail processes. Apartments and condo buildings are adding smart package rooms.

- **Mail/package storage:** Smart storage apps offer contactless delivery, safekeeping, delivery, and pickup for staff and tenants. Tracking and notification solutions are integrated with the lockers. 24/7 access via an app allows for convenient safe pickup. Preconfigured or customizable lockers are available to accommodate package sizes and quantities in real time. Using the smart package delivery app, occupants can command robots or drones to deliver the package upon request. Smart storage delivery is often cited as the number one amenity for enhancing the occupant's experience.

- **Manufacturing and industrial:** IoT sensors are used for supply chain logistics. These sensors track objects and measure and report via various apps the objects' vital conditions, such as temperature, weight, pressure, and other conditions. They track the route and current location and provide remote production and quality control.

- **Metering:** Smart metering can monitor and measure consumption for gas, electricity, and water to ensure each tenant is billed according to their actual use. Using a smart metering app, tenants can implement and measure conservation practices. Building operators can simplify their activities by transferring the billing, payment, and processing for each tenant directly to the utility company and delivering invoices via an app. Smart metering can also detect and alert problems with irregular use, such as a water leak that, if gone unnoticed, might have caused other damage throughout the building.

- **Monitoring/Virtual Network Operations Center (VNOC):** IoT technology and customized apps help building operators manage their buildings from a distance. Significant efficiencies and full asset visibility can be achieved along with real-time alerts to provide an overview of information for decision-making. Almost any physical device can be fitted with an IoT sensor and monitored remotely.

Smart remote sensors are helping facilities managers to keep properties in good working order and free up engineers to then take on more complex tasks. Other sensors can flag risks, such as burglaries, gas leaks, or poor indoor air quality. Remote dashboard monitoring delivers real-time data to identify issues at a much earlier stage. Some systems use machine learning to automatically adjust and redefine maintenance plans to prolong the life of the appliances.

- **Movement of building**: **Structural health monitoring** (**SHM**) sensors such as vibration, three-axis deflection, strain, and stress IoT sensors detect movement and seismic activity and provide real-time alerts and reports via an app.

- **Occupancy sensing**: Smart sensors are deployed around the building to detect an individual's presence. They then either cause a reaction locally such as turning on the lights, or they can send real-time data to a centralized management system that may adjust the temperature and ventilation system. These sensors, when coupled with an interactive app, are used for real-time space planning and management decisions.

- **Operations and Maintenance** (**O&M**): O&M is the day-to-day activity necessary for the building, its systems, equipment, and occupants to perform their functions. O&M includes the maintenance of the physical building itself, management of the building's systems, landscaping, groundskeeping, site improvements, and maintenance of the furniture, equipment, and the building's interior. IoT sensors and smart apps are used to track and report real-time results.

- **Parking**: IoT sensors and smart parking apps are used to track and communicate open parking spaces. There are three types of sensors used: ground sensors, overhead or camera-based sensor technology, and simple counter systems that count the number of vehicles and open and close the gate based on availability.

- **Power evolution**: Using specific apps to manage these power networks, **Power over Ethernet** (**PoE**) and **Power over Fiber** (**PoF**) are becoming popular solutions to deliver power to IoT networks, sensors, and devices. Savings are achieved by eliminating separate power supply cabling and outlets.

- **Predictive maintenance** (**PdM**): Predictive maintenance apps are designed to identify potential maintenance issues before they become a problem. With the rise of smart sensors and IoT, these apps make maintenance smarter, cheaper, and more efficient. The sensors are installed on or near radiators, boilers, pumps, and other machinery. They detect critical levels of noise, vibration levels, leaks, or changes in temperature, and when a certain threshold is achieved, the smart system automatically orders the part and schedules the repair before the issue escalates into a system failure.

- **Property management**: IoT sensors can be used with property management software apps to automate dozens of property management tasks. Property management apps allow building operators to track leases and work orders, collect rent, and manage operation finances. Utility metering/submetering, for example, ensures accurate monthly billing. There are hundreds of smart property management apps available these days.

- **Public safety systems**: These in-building amplification systems are commonly referred to as public safety **distributed antenna systems (DASs)**, public safety **bi-directional amplifiers (BDAs)**, or public safety repeaters. IoT sensor solutions including Bluetooth along with AI apps are helping public safety personnel make better decisions. Audio, video, and other IoT sensory data from smart devices provide emergency managers and first responders with real-time awareness and data used for policy decisions, training, and other improvement initiatives.

  Proximity and occupancy IoT sensors can help determine locations for individuals and first responders, and wayfinding applications can guide first responders in and occupants out. Field communications capabilities are enhanced with smartphones, laptops, and tablets, as well as the use of facial recognition and fingerprint sensors. These smart public safety apps and technologies create transparency, improve communications, and help the building occupants be safer.

- **Refrigeration**: Walk-in coolers/freezers and other refrigeration apps use IoT sensors to measure ambient temperature, ultra-low temperature, and high-temperature probes and produce automatic temperature log reports.

- **Resource usage**: Resource use can be optimized with IoT and smart building apps to support a distributed workforce by adjusting physical space for workgroups, tracking assets, monitoring consumption, and delivering a communications network to support mobility throughout the building.

- **Restroom, smart**: IoT touch-free sinks, paper towel dispensers, and toilets also alert building maintenance via an app when soap, toilet paper, and paper towels are running low. Indoor air quality sensors, leak detection sensors, and occupancy sensors can alert janitorial and maintenance teams when service is required.

- **Safety and security**: IoT sensors, cameras, smart devices, beacons, tags, and even lighting all play a role in the building's security processes. Access control systems and visitor registration systems are two highly important parts of the overall building security system. Security cameras and video surveillance inside and outside of the building connect to the IoT network.

  Sensors and door alarm systems can activate loud noises to deter criminal activity while in progress. Smoke and fire sensors along with fire suppressant systems use IoT sensors to meet code requirements. Commercial cybersecurity systems protect the building's networks and typically include antivirus software, traffic monitoring, data encryption, and firewall protection. Gunshot detection sensors can be coupled with emergency notifications.

- **Space management**: Social distancing guidelines, hybrid-working arrangements, and continuous virus-variant outbreaks have building space planners searching for technology solutions. By combining IoT sensor data with advanced analytic solutions and space management software, they can deliver actionable solutions in many areas to save money and enhance daily building operations.

  Workspace changes can be made in real time by analyzing vacancy rates, utilization rates, and usage behavior. Occupancy sensing and space utilization information can provide the ability to forecast space requirements. Space planning and utilization can be optimized using real-time IoT-driven heat maps and employee work patterns. Underutilized space can be cut back while high-use space can be better equipped and expanded.

Individual space versus open space requirements can be optimized using the data collected and room capacity can be tracked for better utilization. Space planning and reconfiguration information can be used for changes in lease contracts and flexible seating. Space management software can help determine which employees need to be onsite and which could work remotely. Desk assignments, hoteling, or hot-desking options can be managed in real time.

- **Submetering**: Submetering places IoT meters on each floor, for each utility, and for each individual unit. This allows for accurate readings and billings for each tenant for actual utilities they consumed. Tenants can track their usage and bills can be delivered using a smart app.

- **Tenant controls and applications**: The **Tenant Mobile Application (TMA)** is an app specifically designed to give tenants more control over their environment. The app also will digitalize tenant-facing services, transactions, and work orders, to enhance the overall **Quality of Experience (QoE)** for the building's occupants with numerous amenity services. Tenants may directly control lighting, temperature, humidity, and other comfort parameters for personalized comfort and convenience.

- **Ventilation**: Like HVAC IoT solutions, IoT sensors are placed directly in the ventilation system to monitor and measure normal systems operations. Data is sent via the IoT network to a local computing platform and analyzed using an app. As new data is collected and transmitted, it is compared to the normal data to determine whether any abnormal operations and adjustments are automatically made when required. **Demand-Controlled Ventilation (DCV)** uses signals from air-pollutant sensors to regulate the ventilation airflow.

- **Virtual reality (VR)**: Building operators are using VR apps to conduct visual walkthroughs to spot potential problems, visualize space planning, and perform dynamic quality checks. They can better associate the data provided to them with the asset directly. VR training apps deliver on-the-job training using real-life settings through an immersive learning experience to train building technicians.

- **Water quality**: A **smart water quality monitoring (SWQM)** app is an IoT-based system used to measure water quality parameters, such as pH, temperature, oxygen levels, variety of ions present, turbidity, and so on. Various IoT sensors replace the physical method of monitoring with chemicals. These sensors may also contain a specific sensor for leak detection and flow monitoring.

- **Wayfinding and navigation**: Like outdoor-based turn-by-turn GPS map guidance systems, indoor wayfinding and navigation apps help individuals navigate their way around the building and other unfamiliar environments. It safely manages the movement and flow of people through the building, while encouraging social distancing. It improves the user experience and contributes to a sense of wellbeing and security. **Real-Time Location Systems (RTLSs)** generate live updates on the location of people and objects using sensors and a communications protocol such as Bluetooth.

- **Windows**: Smart windows use embedded or add-on IoT sensors and smart apps to adjust shading and tinting levels throughout the day. These windows are energy efficient and sometimes also act as transparent solar panels. Some windows contain antennas to help bring the cellular network into the building. Each window has a unique **Internet Protocol** (**IP**) address used to identify itself and communicate with other devices. Since they tint automatically or via a remote-control app, no bulky expensive blinds are required, and they control heat and glare. Immersive display windows/glass may also transform into digital, interactive surfaces.

- **Workflow and productivity**: **Workforce Management** (**WFM**) apps combined with IoT-connected networks provide access to critical company data to ensure the workforce is distributed efficiently and effectively. Smartphones, wearables, and sensors collect data used by managers to allocate resources and track staff performance on a task-by-task basis. Effective WFM systems need to be able to collect workforce statistics regarding worker performance in real time. They require the capabilities to analyze the current situation and provide improvement recommendations.

- **Work orders**: Work order apps are used by tenants to submit and track their work requests along with receiving cost options and estimates if required. Digital work order systems allow work orders to be completed on a smartphone. The status of the workflow is updated at each step using IoT, barcodes, QR codes, and other smart sensors.

  Work orders are used to provide real-time status and prioritization, and they provide a digital record that can be used for invoicing, audits, and historical record keeping. By installing IoT sensors throughout the building, worker movements can be tracked, and work orders are updated immediately.

  Beacons and asset tags situated all over the building establish control points that supply data on where the building engineer or repair person has been. QR codes installed in the restrooms can be scanned by the cleaning crew upon completion to date and timestamp the cleaning, while notifications can be sent to management and others.

- **Workstations, adjustable ergonomics**: IoT solutions can transform building ergonomics. Smartwatch apps can provide data to an adjustable standing desk about a person's height and adjust the desk when the person approaches. Chairs with IoT sensors can adjust and notify the person when it's time to take a movement break. Other IoT wearables can notify a person to correct an injury-prone posture.

- **Vibration**: IoT vibration sensors placed on equipment and systems measure abnormal vibrations and changes in temperatures to deliver early warnings of potential system failure. These system failures may be caused by component imbalance, wear, misalignment, or incorrect use of the equipment or system.

While this list is lengthy, it most likely is not exhaustive. New technologies and applications continue to be introduced, and smart building planners should remain connected to industry-related websites, conferences, newsletters, and other available sources of up-to-date information and trends.

IoT middleware

IoT middleware is software that is the middle communication layer between different components, devices, systems, and other building communication items. This software creates an interface and sits between IoT devices and other devices that traditionally would not be capable of communicating. It helps legacy systems communicate and to find common ground with newer technologies. It performs like an interface between smart devices helping them to communicate. RedHat, MuleSoft, Oracle, and WSO2 are some of the companies providing cloud-based IoT middleware products to speed up and deliver proper connectivity.

Since the goal of IoT in smart buildings is to make everything capable of connecting and communicating across a network, middleware enables large numbers of IoT devices to be implemented by providing a connectivity layer for the sensors to connect with the application layer. This software performs API management, along with basic routing, messaging, and message transformation functions.

Middleware helps engineers easily build integration points by replacing dozens of unnecessary tools and connections. Its clear architecture ensures seamless integrations to provide smooth data traffic and uninterrupted operations. It reduces the integration time and lowers the total cost of ownership. These solutions are increasingly flexible, agile, scalable, and adaptable. Middleware also adds an additional layer of security since it provides authentication and authorization requirements to access connected devices.

Codes, standards, and guidelines to consider

As the building industry evolves from intelligent buildings to smart buildings, new vendors join the marketplace indicating they have the best solution yet. Existing vendors continue to try to protect their foothold with their proprietary solutions vendor-locking others out and holding building owners hostage.

To prevent owners from becoming pigeonholed into one solution, they should consider selecting solutions that are open sourced and use open standards and codes. This allows for over-the-top solutions to be integrated into the network by sharing common industry protocols. There are several codes, standards, and guidelines that should be considered when selecting smart building hardware and applications solutions. They are as follows:

- **ANSI/BICSI-007-2020, Information Communication Technology Design and Implementation Practices for Intelligent Buildings and Premises**: This is considered the fundamental standard used in designing and implementing **information communication technology (ICT)** infrastructure in buildings. This standard supports network-enabled building systems including traditional, IoT, and smart networks.

The standard is continually updated to include the latest methods and trends for building systems, cabling, reliability, design, and security. It supports the network infrastructure, the materials used for the common building systems, and the methods for powering these systems. Other major items covered include the following:

- BASs
- BMSs
- Low voltage/PoE lighting
- Access control
- **Energy management systems (EMSs)**
- Asset management (RFID)
- **Electronic security systems (ESS)**
- Vertical transportation (for example, elevators)
- Video surveillance
- Intrusion detection
- Digital signage and wayfinding
- Fire alarm
- Intercom, paging, and mass notification
- Sound masking

- **ASHRAE Standard 135 BACnet – A Data Communication Protocol for Building Automation and Control Networks**: This protocol outlines the method by which computerized equipment exchanges information between the various building systems. It was designed specifically for buildings and supports wired, wireless, and cloud-hosted devices.

The protocol outlines a set of messages for sharing encoded data between building automation devices including, but not limited to, the following:

- Hardware binary input and output values
- Hardware analog input and output values
- Software data values
- Schedule information
- Alarm and event information
- Trend and event logs
- Files

- Control logic

- Application-specific data for a large range of building services

- Network configuration including security

- **Brick Schema Uniform Metadata Schema for Buildings**: This open source initiative standardizes descriptions of the logical, physical, and virtual building assets and the interactions between them. Brick contains a dictionary of concepts and terms for elements in and around the building. It defines relationships for composing and linking concepts together. Basically, it establishes common metadata so that the building's data means the same thing and is understood by people not involved in the collection and creation of the data.

- **EN ISO 16484-1 Building Automation and Control Systems**: This is a European standard that outlines guiding principles for designing and implementing other systems into the **building automation and control system** (**BACS**). The standard specifies the phases for the BACS project, including design, engineering, installation, and completion.

- **ISO/IEC 18598-2016: This standard is for automated infrastructure management** (AIM) systems and covers requirements, data exchange, and applications. Technology agnostic, the standard covers integrated hardware and software systems to do the following:

  - Automatically detect when cords are inserted or removed

  - Provide documentation for cabling infrastructure and connected equipment

  - Provide instructions and data management and exchange with other systems

- **Project Haystack**: These are open-sourced standardized semantic data models and web services that streamline working with IoT data. They are used to gain insights into the data generated by smart devices and include applications such as energy, automation, building controls, HVAC, lighting, and other systems. Building owners and other stakeholders can specify Haystack conventions to be used to position their systems for the future of other added services.

- **Telecommunications Industry Association (TIA) smart buildings program**: This program is comprised of over 60 member companies that develop industry standards and certification programs to help building owners measure their building's current intelligence level and provide guidance for future technology developments.

In addition to these codes, standards, and guidelines, attention should also be paid to cybersecurity and network security standards and guidelines.

Summary

The major hardware components necessary to build a smart building IoT network include IoT sensors, connectivity and communications devices, computing capabilities either locally, on the edge, or in the cloud, output devices, and user displays. Smart applications are then required to deliver the solution, and this chapter provided an extensive list of applications that make buildings smarter. IoT middleware is software that is the middle communication layer between different components, devices, systems, and other building communication items. We concluded by recommending that planners review key industry codes, standards, and guidelines to ensure their smart projects do not quickly become obsolete.

The next chapter will focus on how to construct and integrate your smart building project in a new construction or retrofit environment, along with identifying the service providers that will be required.

Part 4:
Building Sustainability for Contribution to Smart Cities

This part aims to provide a roadmap for how to build and maintain your smart building project. It discusses what partners must be engaged and the impacts and challenges on smart cities and sustainability initiatives.

This part contains the following chapters:

- *Chapter 12, A Roadmap to Your Smart Building Will Require Partners*
- *Chapter 13, The Importance of Smart Buildings for Sustainability and the Environment*
- *Chapter 14, Smart Buildings Lead to Smart Cities*
- *Chapter 15, Smart Buildings on the Bleeding Edge*

12

A Roadmap to Your Smart Building Will Require Partners

A common misconception is that simply adding IoT and smart devices to a building is enough to position it as a smart building. As we have discussed in previous chapters, the smart building is a set of connected smart systems integrated and interoperable through a connected platform, capable of analyzing real-time data to deliver actionable insights.

While we have documented the benefits, use cases, components, available technology, and architectures well, the roadmap to achieve a smart building isn't as clear. This chapter offers a roadmap for existing and new buildings to make your building smarter as each new system is integrated. We'll include references to previous chapters to indicate where that chapter's subject matter fits in with the roadmap. We'll identify the various partners that may be required beyond the construction crews.

In this chapter, we're going to do the following:

- Realize the definition and differences between a built environment and new buildings

- Uncover the different types of projects for the built environment

- Discover the various partners and their roles, and see how they are involved in traditional and smart building projects' design, implementation, and integration

- Examine a roadmap to follow to deliver your smart building projects, from design to integration, and provide references to previous chapters to pull together a unified smart project plan

- Become aware of the potential roadmap roadblocks

Smart project planning

Making buildings smarter will be dramatically different between new construction projects and retrofitting an existing building (better known as retrofitting the built environment). Let's begin to define a smart building project by first understanding what type of project environment you are dealing with.

A built environment drives the majority of IoT smart buildings projects

The term *built environment* generally describes an environment that is everything that surrounds us and that is built by humans. In its holistic view, it includes infrastructure, bridges, roads, sidewalks, transportation, buildings, and all the spaces between buildings. For the sake of this discussion, we will use the term to describe existing buildings, regardless of building type and their surrounding spaces.

Of the total number of buildings projected for the year 2050 worldwide, 80 percent of that number already exists today, and they make up the *built environment*. Suffice it to say that the majority of IoT smart building projects will take place with existing buildings.

An existing building is comprised of four major components:

- **The building's exterior**: The exterior is the visible component, which includes windows, doors, walls, and roofs. It not only represents the appearance of the building but also impacts how well the building functions as a shelter. When poorly constructed, there could be significant heat loss and rain penetration.

- **The building's interior**: The interior is where people work, live, and play, and therefore comfort and easy access are the main concerns. It is also the location for the building's systems, such as HVAC, plumbing, and lighting.

- **The structural frame**: This is the foundation that supports the other components of the building, such as walls, floors, and ceilings. It will support considerable amounts of weight and must therefore be designed and constructed properly.

- **The building's finish**: These are the techniques and materials used to create the building itself and its overall appearance.

There are seven other components of the built environment, which include ambient noise, climate, air quality, physical activity, public spaces, land, and use development patterns and transportation.

Smart building projects in the built environment generally range from adding a single smarter solution, such as smart lighting, to a total building retrofit from the ground up. More often, these projects are in the form of a renovation, refurbishment, or system retrofitting:

- **Building renovation**: This is the process of changing an existing structure into a new and better one. It involves bringing something old or damaged back to its original state, and more often than not, the project will involve improvements to make it function better or to make it more aesthetically pleasing. This is an opportune time to add solutions to make the building smarter. For a commercial building, there are generally two types of renovations:

- **Capital improvements**: These projects generally include common areas, such as lobbies, restrooms, and items exterior to the building. These projects typically improve the lives of all the tenants and involve amenity projects, such as fitness centers, conference rooms, laptop bars, and building-wide Wi-Fi. The list of IoT smart building opportunities in this area is very lengthy, as described in *Chapter 5, Tenant Services and Smart Building Amenities*.

- **Tenant improvements**: Commercial offices typically start as a bare-box space without walls, fixtures, and finishes. The tenant will upgrade the space to their own specifications, and this is a good time to work with them on adding smart building features from the get-go. For existing spaces, the building owner will look for opportunities to upgrade to make the space more marketable. Opportunities exist with new IoT lighting, IoT comfort features, and numerous other IoT-related amenities.

- **Building refurbishment**: This is the process of equipping, cleaning, or decorating such as cosmetic improvements, including painting, upgrading, repair work, alterations, extensions, conversions, and modernization. Each of these is an opportunity to add IoT smart building solutions to a project.

While the terms *renovation* and *refurbishment* are often used interchangeably, they are two different classifications in the building industry. The difference is that renovation typically means to restore, and refurbishment generally means to improve.

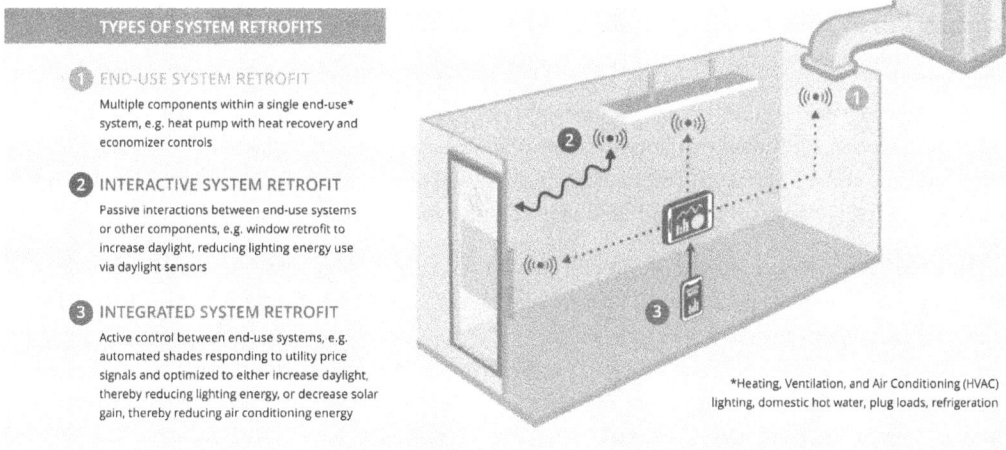

TYPES OF SYSTEM RETROFITS

① END-USE SYSTEM RETROFIT
Multiple components within a single end-use* system, e.g. heat pump with heat recovery and economizer controls

② INTERACTIVE SYSTEM RETROFIT
Passive interactions between end-use systems or other components, e.g. window retrofit to increase daylight, reducing lighting energy use via daylight sensors

③ INTEGRATED SYSTEM RETROFIT
Active control between end-use systems, e.g. automated shades responding to utility price signals and optimized to either increase daylight, thereby reducing lighting energy, or decrease solar gain, thereby reducing air conditioning energy

*Heating, Ventilation, and Air Conditioning (HVAC) lighting, domestic hot water, plug loads, refrigeration

Figure 12.1 – Types of system retrofits (diagram courtesy of the LBNL)

- **System retrofit**: A building system combines equipment, controls, operations, accessories, and connectivity to deliver a particular function or service, such as heating, lighting, and ventilation. Often, these systems can be retrofitted (a system retrofit), and the **Lawrence Berkley National Laboratory (LBNL)** has identified three common types of system retrofits:

  - **End-use system retrofit**: One to multiple components are retrofitted/replaced within a single system, such as adding a new heat pump with IoT heat recovery and economizer controls

  - **Interactive system retrofit**: The passive interactions between components and end-use systems are retrofitted, such as replacing windows with IoT sensors to increase daylight use and reduce energy consumption

  - **Integrated system retrofit**: Integrated and interactive control is added to the end-use system, such as automated shades that respond to signals from utility providers regarding price fluctuations to optimize daylight shading, thereby reducing lighting energy and HVAC energy use

The LBNL conducted research on over 12,000 retrofit projects, with information from utility companies and government programs, and determined that system retrofits are critical in reaching energy-reduction and sustainability goals.

New buildings create built-in IoT opportunities

When building a brand-new building from the ground up in today's environment, consideration should be given to how smart the building will be both when the new build has been completed and beyond. Some new buildings will be built with primary goals to achieve energy reduction, operations efficiency, sustainability, and occupant satisfaction, and they will, therefore, include as much IoT smart building technology as currently available and simultaneously.

Other new buildings will be built with a phased approach in mind, meaning they will add smart IoT solutions over time as additional funding becomes available and as occupant demand dictates. Either scenario requires consideration of what a fully integrated **Information Technology/Operational Technology (IT/OT)** infrastructure will look like well into the future.

Planning a fully integrated smart building will start at the design stage and follow the same smart building project roadmap that is described later in this chapter. **Building Information Modeling (BIM)** uses the latest technology to create 3D models to visualize the functionality and physical characteristics of the building. BIM is used by **Architecture, Engineering, and Construction (AEC)** professionals and more recently by facility managers. It improves asset allocation and provides insights into spatial awareness, light analysis, and the properties of building components.

Augmented Reality (AR) and **Virtual Reality (VR)** are being used to supply environmental elements through computer-generated sensory input, supplied by IoT sensors and cameras. Using tablets, glasses, or even a smartphone, AR can superimpose objects onto a person's view of their physical surroundings. VR creates virtual mock-ups for building owners to experience the designed structure as if they were there long before the first shovel of dirt is moved.

Digital twins are visual representations of a building and are discussed in detail in *Chapter 8, Digital Twins – a Virtual Representation*. Whereas BIM diagrams typically display numerous systems and subsystems while focusing on how the building is constructed, digital twins show space and how people will use it. These digital twins bridge the gap between form and function. They remain useful after construction and are used for the management of the building and to assist first responders in quickly understanding the building's layout, as described in *Chapter 3, First Responders and Building Safety*.

IoT technology is also helping the construction industry during construction, with equipment and tool tracking not only for location but also for repair, maintenance, and scheduling purposes. IoT security sensors and cameras aid in securing a construction site to reduce theft. Using IoT wearable devices and sensors, we can measure construction worker conditions, such as health, performance, and productivity.

Chapter 15, Smart Buildings on the Bleeding Edge, examines IoT in the construction of smart buildings. Construction site safety is a priority, and wearable sensors can notify all parties involved when heavy machinery is approaching a worker. Exoskeleton wearable frames and sensors worn by workers can provide them with additional support, assist them when lifting heavy objects, and disperse weight when needed.

Partners

As smart building projects become increasingly more complex, smart building support partners are brought in to help design and integrate building systems. This complexity also changes the role these support partners play, evolving from conventional product integrators into master systems integrators and consultants. The following is an overview of the different partner roles to consider when determining what you'll need for your smart project:

- **Conventional product integrators**: These engineers are solely focused on incorporating a single product or subsystem into a building and integrating that subsystem into the overall building automation scheme. Typically, they are deeply connected to the **Original Equipment Manufacturer** (**OEM**) supplying the system, and they are highly trained in that system's hardware and proprietary language. Unfortunately, this usually results in disparate, siloed systems that are not integrated.

- **Managed Service Providers** (**MSPs**): MSPs are outsourced companies that take the responsibility of designing and maintaining one or more of the building's subsystems. They are strategic business partners who have the expertise and latest technology to remotely manage a specified system or function.

  Typically, there will be a **Service-Level Agreement** (**SLA**) that defines the expected level of service required and the consequences for failure to deliver. Some of the downsides to using

MSPs are that you'll need several of them to cover a large number of subsystems, and many use proprietary technology and tools, which creates problems for clients after the contract is canceled.

- **Mechanical, Electrical, and Plumbing (MEP) engineer**: There is a high level of interaction between the MEP disciplines, and they are usually addressed together to prevent conflicts in the equipment locations. Computer software is used by MEP consulting firms to speed up the design process by automating repetitive tasks. By having these disciplines work together, buildings can be designed to be more energy-efficient, be more sustainable, and use fewer resources.

MEP engineers work with architects, building owners, and trade contractors from the planning stage through to the post-occupancy survey. They manage the following construction-related documents:

- Detailed diagrams and drawings of each floor plan, along with their elevations and sections

- Technical specifications for each system to be integrated

- Lists of required materials and products

- Execution methods for the materials, products, and systems

When consulting firms combine these three MEP disciplines, projects become more streamlined and synchronized. With the increase in smart buildings and building automation, centralized hardware and software networks allow these systems to work together to reduce delays and confusion.

- **Systems Integrator (SI)**: SIs specify, design, install, and maintain a building's subsystems to deliver an integrated solution for the building. They are individual engineers or a tech company that is contracted to manage the **Building Automation System (BAS)** and other open control systems, with a focus on trends, alerts, and schedules.

They assemble various components from different suppliers to deliver a cohesive structure to meet a building owner's requirements. By using SIs, the building owner can reduce the total number of contractors they need to operate and maintain the subsystems.

- **Master Systems Integrator (MSI)**: MSIs go beyond MEP engineers and SIs, with a stronger field of expertise. They have skills in database engineering, **IT**, software development, and cybersecurity. These skills are combined with traditional subsystem skills, such as energy management, lighting, HVAC, access control, and security systems, and non-traditional fields such as workplace performance, wellness, and environmental quality. MSIs use their collaboration skills to pull together disparate siloed systems into one turnkey solution.

As a key consultant to the building owner, the MSI creates and executes a building's smart strategy and technology roadmap. They bring together IT and OT for the building's network convergence and often develop software solutions to augment systems integration. They perform difficult API integrations, are experienced in database management and custom analytics, and have cybersecurity experience.

They can work with non-standard communication and complex integration projects, perform life cycle infrastructure management, and often have capabilities to monitor systems in real time. The following are some of the roles that an MSI will perform:

- Smart building consultation

- Workflow optimization

- Custom software development

- Data validation

- Data transformation

- OT network design and implementation support

- Enterprise smart sequences of operations

- Non-standard communication protocol support

- Utility bill validation

- Equipment performance monitoring

The key difference between the SI and MSI is the MSI's ability to take the building owner's vision and execute a strategy to deliver the solutions for a smart building integrated network. While the MSI may subcontract many parts to SIs, they have the overall knowledge to coordinate everything.

- **Consultants**: Smart building consultants are less focused on the systems and sensor technology and are focused more on clients' goals. Since buildings are built for people, their belief is that technology should enable a better experience and not complicate it. Owners engage consultants early on in their design process many times, even before any specific project is determined. The consultant guides conceptual vision sessions, develops the foundations that support the integrated systems, and selects the systems to meet the goals.

Regardless of the type of partner(s) brought into the project, it's important that each has a deep understanding of the business, the client, and the goals of the project before thinking about design solutions. They will need to justify the return on investment and the impact on productivity and the occupant's experience.

Smart building project roadmap

Each smart building project will require adding smartness, connectivity, and convergence by bringing together many different components, technologies, and individuals. Since IoT smart integrated buildings are relatively new, there are few industry-designated smart building design certifications and programs and, therefore, a limited number of graduates.

There are, however, experienced architects, engineers, and consultants as we identified earlier who have gained experience through their projects, and they can offer their expertise to help guide you through your project. Whether or not you engage a consultant to assist in your smart project, we recommend following a phased roadmap approach to keep the project disciplined, focused, and aligned. *Figure 12.2* outlines the roadmap phases and aligns previous chapters as references for detailed information.

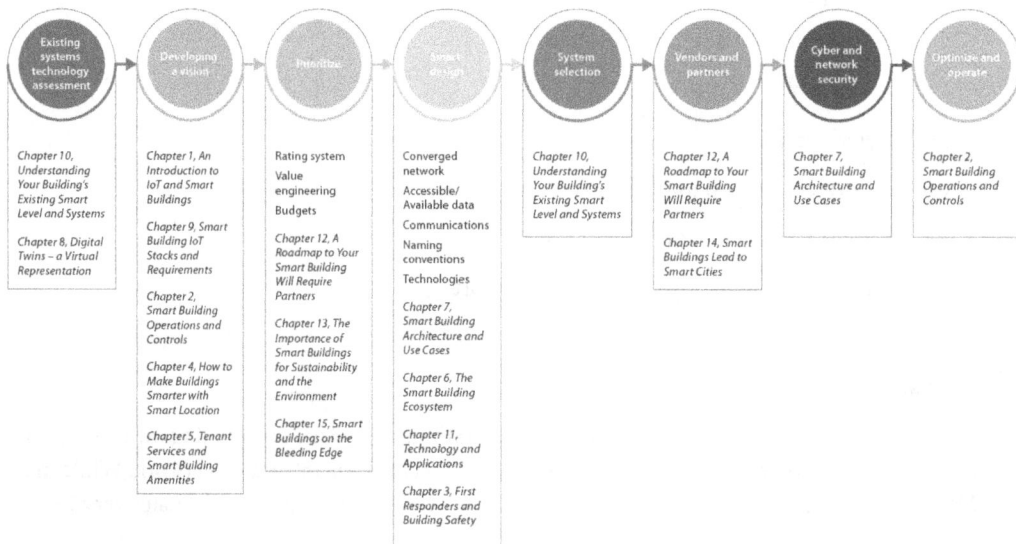

Figure 12.2 – A smart project roadmap and related chapters

These are the smart building project roadmap phases:

1. **Existing systems technology assessment**: In *Chapter 10, Understanding Your Building's Existing Smart Level and Systems,* we identified a detailed list of building systems to be considered as candidates to implement IoT and smart building solutions. This list of existing building systems is the starting point for any smart project.

    Chapter 10, Understanding Your Building's Existing Smart Level and Systems also outlined methodologies to determine what systems are currently in a building and how smart they may already be, by using various available industry assessment processes for an existing building. This assessment considers the computing power, current technologies implemented, the age of the facility and its systems, the ability to integrate with other systems, IoT compatibility, and what types of data you already have access to. The key questions for existing buildings are what is missing, whether these systems can be upgraded, or whether they need to be replaced.

Figure 12.3 – New building smart design begins at the design phase

This assessment could be performed for new construction in a virtual model as well. New buildings are often considered to be an easier smart building project, since we start from scratch, and we can use virtual designs. Smart design must start at the initial infrastructure construction design stage to identify IoT equipment integration requirements. For example, the elevator system is traditionally the first system to be built; therefore, upfront planning is required to ensure that it can be integrated and considered for smart elevator implementation.

2. **Developing a vision**: The term *smart building* has different meanings to different people and involves more than just connecting various devices and systems. The outcomes of a smart building project extend to many stakeholders, including owners, operators, occupants, and visitors. *Chapter 1, An Introduction to IoT and Smart Buildings*, identified a large number of potential stakeholders. It also provided insights into the vast possibilities of smart solutions that contribute to making a building smart, in our example of a day in a smart building.

    The purpose of developing a project vision is to discover and develop together with the stakeholders the current operational issues and the desires for their business as a whole, not just for the building.

    Most of the previous chapters provided examples and use cases of solutions available to reduce operational costs, improve efficiency, reduce energy consumption, and improve an occupant's **Quality of Experience** (**QoE**). For example, *Chapter 5, Tenant Services and Smart Building Amenities*, highlights several smart projects focused on improving occupant satisfaction.

    Key questions to consider when developing a smart project vision include the following:

    *   Who are the project stakeholders?

    *   What is important to each stakeholder?

- How will the stakeholders and users interact with the technology?

- What problems are we trying to solve?

- What value will this vision bring to the owner, operator, and occupants?

- What training will be required to engage the users with the technology and data?

In *Chapter 9, Smart Building IoT Stacks and Requirements*, we provided a methodology to use to develop your smart building requirements. While most building owners are interested in their bottom line, the most common goals are energy and water optimization, occupant or employee productivity, and real estate operational efficiency. These are proportionally related and compound quickly; however, the list will go far beyond these three areas.

Smart asset and people location, indoor positioning, navigation, way-finding, tracking, and occupancy sensing are popular smart building IoT projects and are detailed in *Chapter 4, How to Make Buildings Smarter with Smart Location*.

Smart spaces that make employees happier, more comfortable, and more productive can have a much bigger impact on the overall bottom line than reducing and optimizing energy use. The space management section in *Chapter 2, Smart Building Operations and Controls*, discusses the benefits of smart space planning.

Not every project has a direct monetary return – for example, the outcome may only be useful data to support other goals and initiatives. Specific metrics are defined during this visioning phase and are used to measure performance throughout the life of a business. Questions can include the following:

- What do we want to achieve with our smart building project?

- What does the return on investment look like?

- What metrics should be measured and improved on?

- What benchmarking information do we have already and what new benchmarks may need to be developed?

- Is this an all-at-once project or a phased rollout?

- Is it a goal to minimize the number of software packages by centralizing operations?

- What systems can we leverage already and what data is currently available?

Financial modeling of metrics is used to quantify the owners' goals. These may include projected energy savings, occupant comfort metrics, and operational efficiency calculations. By modeling, smart projects that appear expensive at first can demonstrate their return value, even when using conservative estimates.

3. **Prioritize**: Unless you are given an unlimited budget (congratulations if you are), you will need to prioritize projects by taking the building assessment and requirements identified earlier and creating a rating scale to propose to the decision-makers. A simple grading system from 1 to

10 can be used to rate each piece of equipment or system to be considered for upgrade, and for each new smart solution to be considered.

From my perspective, the worst words I have heard in this industry are **value engineering**. This is supposed to be a systematic and organized approach to finding ways to deliver a desired functionality at the lowest cost possible, by substituting materials and methods while not impacting functionality. Unfortunately, when applied, it usually results in projects being cut from the master plan before they ever have a chance. Therefore, while trying to adhere closely to value engineering's intent, you should prioritize projects in preparation for some being eliminated at the budgeting stage.

Budget considerations and technology maturity levels should be considered to determine whether an investment is worthwhile. Two other issues that need to be considered are facility interruptions and the time of year that the projects will be conducted.

Since most building projects are renovations, refurbishments, and system retrofits, we recommend developing a 5-year smart building plan, with consideration of the technological costs on a traditional bell curve. For example, some technologies may be less expensive today, while other technologies will become less expensive as they mature and gain acceptance.

4. **Smart design**: The key to a smart design is the ability to share data between building systems. At the design stage, theoretical discussions become reality. Vendor in-depth technology reviews are conducted for their ability to deliver compatibility and their smart building solution claims. New technologies are not always what they claim to be.

    The smart design will require considerations for the following:

    *   **Converged networks**: A converged network will centralize the management of the **IT/OT** networks and reduce the amount of hardware required, as discussed in the IT/OT convergence section of *Chapter 7, Smart Building Architecture and Use Cases*. This will create a cost-effective integrated network for the efficient analysis of data.

    *   **Accessible and available data**: Traditional building systems such as mechanical, electrical, and plumbing can each capture large amounts of data. How this data is mined and made to deliver actionable insights will depend on the goals defined in the preceding vision phase.

        Regardless, data should not be stored within siloed systems, and it needs to be accessible and available. While the solution will differ with each project, a single database should be identified to deliver data integrity and accuracy. Other data-related questions include the following:

        *   Who owns the data and who has access to it?

        *   Do you prefer on-premises software or the cloud?

        *   Is there a central data warehouse, or will the data be stored in separate platforms?

        *   What tools are available or will be required to perform data analytics, management, and visualization, as outlined in *Chapter 6, The Smart Building Ecosystem*?

- **Communications**: Which communication protocols will these systems and devices use (that is, BACnet, TCP/IP, Modbus, LonWorks, and so on), and are these interoperable with open or closed protocols? *Chapter 6, The Smart Building Ecosystem*, outlines the most common wired and wireless connectivity and communications protocols used in smart buildings. Our recommendation is to standardize one protocol for all your smart projects.

- **Naming conventions**: In *Chapter 11, Technology and Applications*, we discussed the different naming conventions available in the industry, and we recommend standardizing a naming convention for a smart project.

- **Technologies**: Many decisions regarding what technologies will be used in the project will need to be made, regarding IoT sensors and device types, connectivity, and transmission methods, including wire, cable, Ethernet cable, Power over Ethernet, fiber, and wireless, such as Wi-Fi, Bluetooth, Zigbee, LoRaWan, NFC, NB-IoT, RFID, and cellular.

  How and where computing capabilities will be performed needs to be decided to determine what will be done locally within a building using edge and fog devices, and what will be done using cloud-based computing.

  Software and APIs will need to be chosen along with deciding what smart applications will be deployed.

  All the technologies mentioned in this section are defined in detail in *Chapter 6, The Smart Building Ecosystem*, with additional hardware technology identified, and a lengthy list of the aforementioned applications is available in *Chapter 11, Technology and Applications*.

5. **System selection**: Many of the smart systems that enable smart building solutions are already required by local or state building codes. For example, most energy codes already require occupancy sensors to be used to turn off lights in empty spaces.

   Choosing the best integrated building systems will require considerations of cost, control, flexibility, real-time metrics, meaningful feedback, common tagging, and common communication protocols. These systems should be designed to be flexible to meet future needs. A list of building systems along with smart project examples are included in *Chapter 10, Understanding Your Building's Existing Smart Level and Systems*.

6. **Vendors and partners**: Once the systems are selected, the project manager will need to determine what vendors will be selected and what partners, such as MEPs, MSPs, SIs, MSIs, and engineers, will be used. A description of these potential partners was provided earlier in this chapter. Vendor management will be a key requirement for any project, and the following questions will need to be addressed:

   - Will operations, maintenance, and services be outsourced?

   - How will you compare vendor performance and what **SLAs** need to be implemented?

   - How do you manage contract budgets when changes occur?

7. **Cybersecurity and network security**: When integrating disparate networks, security protocols will be a major concern to prevent cyberattacks and security breaches, as discussed at the end of *Chapter 7, Smart Building Architecture and Use Cases*. Questions to address include the following:

    *   What are the network and security requirements?

    *   Who will manage internet connections used for remote access?

    *   What agreements will be needed from each site to be involved in the network?

    *   What security protocols and best practices will be used?

8. **Optimize and operate**: Upon completion of the smart project, insight dashboards and applications will provide data regarding operations and performance. *Chapter 2, Smart Building Operations and Controls*, highlighted the importance of this information needed for facility controls, building operations, and maintenance. From this data, systems can continually be optimized using analytics, **Machine Learning** (**ML**), and **Artificial Intelligence** (**AI**). To ensure a building is functioning as it was designed and built, several competencies, services, processes, and tools are required for day-to-day operations. An effective maintenance process ensures that critical assets are in good condition, resources are allocated efficiently, procedures/schedules are enforced, and building performance is managed.

    Operations and Maintenance (**O&M**), or operational maintenance as it is commonly referred to, are the day-to-day activities necessary for a building, its systems, equipment, and occupants to perform their functions. O&M includes the maintenance of the physical building itself, management of the buildings' systems, landscaping, groundskeeping, site improvements, and maintenance of the furniture, equipment, and the building's interior.

Potential roadblocks

While our smart building roadmap will help guide you toward your smart project(s), every road, unfortunately, has roadblocks, and identifying and handling these early in a process will help ensure success. These roadblocks include the following:

*   **Lack of an internal smart building champion**: Since a smart building requires cutting-edge technology that impacts the day-to-day functionality of the building, it is often difficult to find an internal champion to see the vision and drive the change while maintaining current operation levels. The internal champion will need to embrace the vision, build the team, clear the budget roadblocks, field the contract issues, and continually push forward to achieve the integration vision. This champion must also have insight into the buildings' mission, energy consumption, security, operations, occupant comfort, and overall productivity. This internal champion is the key to reaching or falling short of smart building goals.

*   **IT and cybersecurity**: The IT and OT network convergence required to deliver an integrated smart building means combining IoT devices on the same network as the company's enterprise network. This means combining two very distinct groups and systems to be cost-effective, but

that may not be ideal in the real world. The challenge is the cybersecurity risk of using a single network. Some buildings opt for dedicated networks while others address the cybersecurity risk head on, with processes and procedures to minimize the risk to their network.

- **Prior decisions**: For both new and existing buildings, prior decisions will impact how successful the IoT and smart building implementation will be. Space considerations for additional equipment, cabling infrastructure, power, and metering equipment could impact the implementation of additional equipment.

  Return on investment models from previous technology deployments may prevent new equipment from being financially viable, since these previous returns may not have been achieved yet. Past technology choices may prevent the implementation of newer desired choices.

Obviously, there are numerous other potential roadblocks, such as newly implemented code and government requirements driven by changing sustainability targets. There is a lack of IoT and smart building skills available to design and complete projects. As we learned with the recent pandemic, potential supply chain issues can arise at any moment, preventing required infrastructure technology from being available and cost-effective.

Summary

Leading a smart building IoT project requires a disciplined approach, knowledge of the industry and its common protocols, and the ability to recognize opportunities. Design, engineering, and technology partners will be required to supplement skill sets, drive implementation, and ensure complete system integration. While new buildings under development are excellent candidates to implement and integrate every latest smart building technology, often budget restrictions will require a phased approach, which means a long-term plan will be required to ensure future technology can easily be integrated.

Part of a disciplined approach includes following a roadmap with defined steps, including assessing existing systems, developing a smart vision, prioritizing projects, designing smart, engaging vendors and partners, addressing security risks, and developing steps to optimize and operate these systems in the long term. In this chapter, we outlined a recommended roadmap and tied previous chapters to each of the phases to help you build your smart building IoT project plan. Every roadmap will inevitably come across roadblocks, and this chapter identified the common ones to look out for.

The next chapter will focus on why and how smart buildings impact the environment and sustainability goals. Buildings consume 40 percent of the energy available, and therefore, smart energy conservation projects can reduce consumption and carbon footprints, foster sustainability, and endorse eco-friendly alternatives.

13

The Importance of Smart Buildings for Sustainability and the Environment

Buildings contribute to over 40% of the world's carbon emissions, and this raises awareness of their impact on the environment, society, and life. The **environmental, social, and governance (ESG)** framework has become an integral part of many building owners' strategic plans but also remains a mystery to most. Investors, bankers, and VC firms offer preferential treatment and easier credit terms to organizations that embrace ESG frameworks.

IoT helps smart buildings to share information, manage operations, and enrich human interaction. Traditional buildings consume substantial amounts of energy to operate, while smart buildings are equipped to better manage energy usage. Sustainable buildings drive economic and social development while protecting the environment. IoT smart building solutions can reduce their carbon footprint, foster sustainability, and endorse eco-friendly alternatives.

In this chapter, we're going to do the following:

- Recognize the difference and similarities between ESG and sustainability as we define and compare each in detail

- Define the common ESG terminology

- Review how IoT and smart buildings help deliver, measure, and report ESG data

- Understand the six principles of sustainability designs

- Discover what green buildings are and the various types of green buildings

- Learn the European Union sustainability taxonomy

- Examine trends in ESG and smart building investments
- Look at ways to fund sustainability projects
- Understand the impact of the Paris Agreement's global sustainability plan

ESG

Climate change, carbon emissions, pollution, deforestation, and other environmental concerns are top-of-mind business concerns for building owners, investors, operators, and other stakeholders. **ESG** is fundamental to both an organization's governance framework and its investment framework.

Organizations define their mission, vision, strategy, values, and tactics with consideration of ESG impacts. Investors and stakeholders look to building owners to understand how they are responding to trends related to climate, demographic, and technological changes as part of the investment decision-making process.

Figure 13.1 – ESG

The ESG acronym was first used in 2006 by the United Nations in their **Principles for Responsible Investing (PRI)** report. At that time, the ESG criteria became part of the financial evaluation of companies and, subsequently, buildings. ESG is sometimes referred to as *impact investing*. It was developed from the assumption that an organization's financial performance is directly affected by environmental and social factors. The belief is that organizations with better ESG scores will perform better, and numerous studies have been conducted to support this.

Here's an overview of the ESG framework in general:

- **Environmental**: Environmental factors include how building owners manage natural resources, direct and indirect greenhouse gas emissions, and resiliency against climate risks, such as climate change, fires, and flooding.

- **Social**: This pillar refers to the building owner's relationship with stakeholders. It focuses on **human capital management (HCM)** metrics such as employment agreements and fair wages, along with buildings' impact on the community. This also includes supply chain partners. It includes workplace health and safety, human rights, government, and community relations.

- **Governance**: Corporate governance focuses on how building owners lead and manage and how their goals align with shareholders' expectations and rights. Internal controls to promote leadership accountability and transparency, executive compensation, accounting, and disclosure practices are examined.

ESG helps organizations manage the risks and opportunities resulting from changing economic, environmental, and social systems. ESG evolved over time beginning as a regulatory framework focused on sustainability. In the 1980s, **environmental, health, and safety (EHS)** programs focused on how regulations could manage reducing pollution to spur economic growth, with an additional focus on labor and safety standards.

In the 1990s, this evolved into a corporate sustainability movement with leaders focused on reducing their environmental impact beyond government mandates. Many accused management teams of *greenwashing*, a practice of using sustainability as a marketing tool, and misrepresenting by overstating their program's real environmental impact.

In the 2000s, the corporate sustainability movement added consideration for social issues and became known as **Corporate Social Responsibility (CSR)**. CSR added employee volunteerism and corporate philanthropy as key components while many critics argued that philanthropy was merely a tax incentive that made tax contributions attractive.

In the late 2010s and early 2020s, ESG continued to evolve and has become a more proactive movement instead of reactive. Key elements of the framework focus on environmental and social impacts while maximizing stakeholder well-being.

ESG terminology

There are a few common terms used when discussing ESG:

- **ESG investing**: ESG was developed by financial institutions looking to make an investment. They understood that an organization's financial performance is often tied to the organization's environmental and social responsibility.

- **ESG metrics**: These are ESG-specific metrics used to assess an organization's exposure to ESG risks. They are used for benchmarking and scenario analyses. Many of the metrics are like traditional financial metrics; however, non-financial metrics such as waste reduction, green gas emissions, water and energy usage, and health and safety incidents are added. IoT-collected data and smart building analysis capabilities are fundamental to providing these metrics.

- **ESG frameworks**: These ESG frameworks are the processes for standardized disclosure and reporting of the ESG metrics. The **Global Reporting Initiative** (**GRI**) is a set of standards for environmental, social, economic, and governance conduct.

- **ESG reporting**: This reporting is the disclosure of the organization's operations performance across ESG areas.

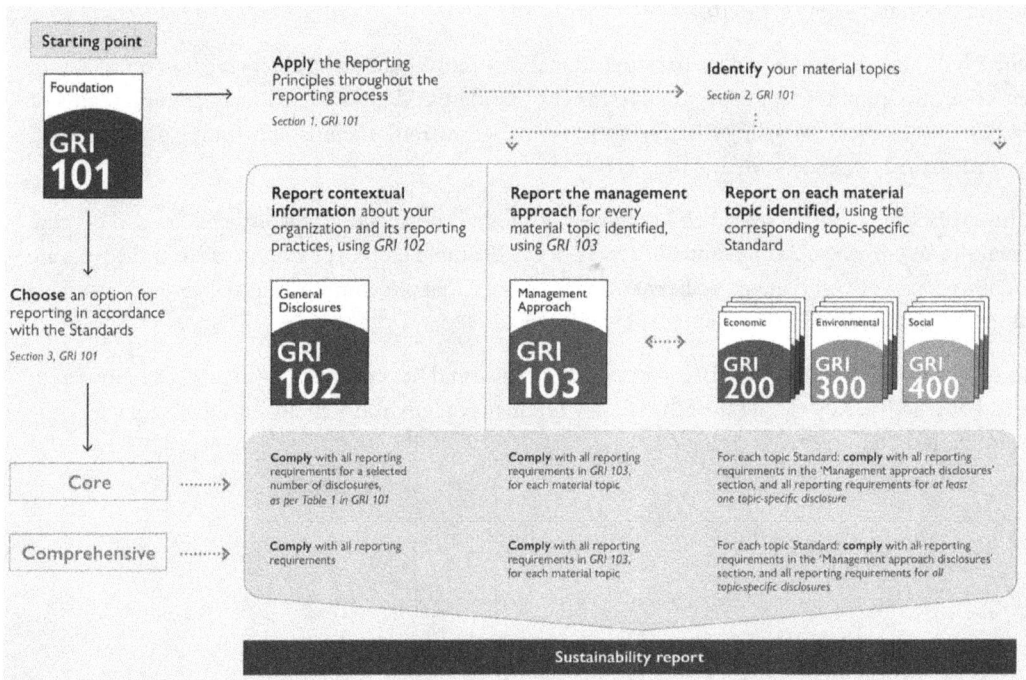

Figure 13.2 – GRI sustainability reporting (courtesy of Global Reporting)

Sustainability reporting using the GRI standards allows multiple stakeholders with differing backgrounds to read and understand the information consistently. Building engineers, investors, financial personnel, tenants, and many others can review plans and results in a comprehensive format.

ESG and buildings

The United Nations' **Intergovernmental Panel on Climate Change** (**IPCC**), among other initiatives, has sounded the alarm on climate change driving people's awareness toward a building's impact on the environment. The 2020 pandemic has brought additional awareness of the health and well-being of a building's occupants.

Most larger buildings (over 50,00 sq. ft) have IoT-driven building automation systems, which help considerably in allowing the building to operate efficiently and produce the required reports on how the building is meeting ESG requirements.

The USA's Department of Energy's **Commercial Buildings Energy Consumption Survey** (**CBECS**) reports that 94% of commercial buildings are small (less than 50,000 sq. ft.) and only 11% of these have building automation systems. The conclusion here is that most buildings are running inefficiently and do not provide energy and carbon data.

Smart buildings use IoT sensors and devices to monitor building operations. The following are some solutions to implement to deliver positive ESG results:

- **Environmental solutions**: The carbon footprint impact from buildings involves the building's orientation and construction materials, energy, water, and waste management. This includes the windows-to-walls ratio, the use of construction material considered friendly to the environment, natural ventilation, and the use of energy-saving systems and equipment.

  Implementing smart building management systems and IoT devices has a significant positive impact on saving water and electricity. Environmental waste management systems and procedures can reduce the overall carbon footprint of greenhouse gas emissions:

  - **Energy savings**: Many of the IoT energy optimization solutions we have discussed in previous chapters, such as IoT motion sensors to turn lights on and off, temperature sensors to control air conditioning, and contact sensors to monitor and manage other devices, are contributors to delivering energy savings. IoT window shading, solar panels, and building energy management systems are other examples. Retrofitting **heating, ventilation, and air conditioning** (**HVAC**) systems by adding ambient environment sensors, smart energy meters, and I/O cards enables monitoring and intervention to eliminate unnecessary runtime. Energy management solutions use data to better understand and manage energy consumption and demand.

  - **Water savings**: IoT leak detection probes and sensors, self-closing taps, and other IoT water management systems can detect unusual water consumption. IoT-driven stormwater management systems can recycle water to be used for water features, gardening, and landscaping.

- **Indoor air quality**: Indoor air quality and CO_2 monitoring with IoT sensors have a positive impact on productivity, moisture control, and other environmental elements. These sensors can immediately detect and report pollutants, harmful particulates, and hazardous gases to keep everyone safer and healthier.

- **Physical security**: IoT systems such as retinal or biometric scanning, QR code readers, and other IoT access systems can ensure that only authorized individuals are allowed on the premises.

- **Waste management**: IoT sensors can measure and analyze waste containers' tonnage and fill levels to determine the optimal hauler service to ensure containers are not emptied too early. Waste elimination and recycling opportunities are examined to ensure regulatory compliance.

- **Resource conservation**: Energy, water, gas, and other resources use IoT sensors to monitor, manage, and gauge usage and leaks. Other smart building solutions manage resource operations and efficiency. To manage greenhouse gas emissions, IoT sensors measure carbon dioxide (CO_2), **nitrous oxide (N_2O)**, methane, **nitrogen trifluoride (NF_3)**, as well as the three types of **fluorinated gases (F-gases)**, such as **sulfur hexafluoride (SF_6)**, **hydrofluorocarbons (HFCs)**, and **perfluorocarbons (PFCs)** to reduce carbon footprints.

- **Others**: The effects on the local ecosystem, the impact on the outdoor environment, and many more criteria are included in the environmental portion.

- **Social solutions**: Buildings have a direct impact on their users and these focus programs have a positive impact on improving social ESG ratings. These programs may relate to workforce diversity, equal opportunity, health, safety, and **people with disabilities (PwDs)** initiatives. Other examples include wellness programs such as fitness centers, community events, commuter benefits for public transportation, carpooling, health benefits, and tuition assistance. IoT smart building solutions contribute to delivering many of these amenity programs.

- **Governance solution**: Governance around building management entails trust, transparency, ethics, and sustainable and structured operating practices. IoT smart building solutions can drive best practices and processes with apps or web-based solutions. Examples include policies on ethics, corruption, environmental operations, compliance, transparency, and accountability. This also includes the governance policies of investors, vendors, and other associates.

Investors with a social consciousness will use the ESG framework to screen potential investment opportunities. A 2020 survey from Fitch Ratings found that 67% of banks used ESG risks to screen their loan portfolios. They found that organizations with low ESG standards were at greater risk, and they found the same was true for buildings with below-average ESG standards.

Many of the IoT systems that buildings implement also have community-wide impacts, such as IoT-based power grids. These grids can manage energy in real time across buildings to detect surges before outages or downtime occur and optimize energy usage by controlling end devices during peak loads.

Sustainability

Traditionally, the erection of a new building has harmed the environment due to its energy use for heating, cooling, lighting, ventilation, and other systems. The building environment accounts for 40% of global energy use and an estimated 30% of greenhouse gas emissions. Recently, building architects and developers have focused more on sustainable building practices being energy-efficient and eco-friendly, reducing pollution, and reducing resources.

There are three pillars to the sustainability framework, often referred to as the **3 Ps of Sustainability**:

- **Planet**: This pillar focuses on reducing and eliminating greenhouse gas emissions and protecting the planet's resources. ESG's environmental pillar includes waste management, eliminating toxic hazards, recycling materials, green energy, and reducing carbon emissions.

- **People**: This pillar is like ESG's social pillar; this focuses on business practices to promote health, well-being, and safety. It includes diverse organizational structures, fair wages, gender equity, and employee welfare.

- **Profit**: The focus here is on the financial viability, transparency, and governance of the organization. Like the economic pillar of ESG, it involves business models, practices, and policies related to employment and economic opportunities.

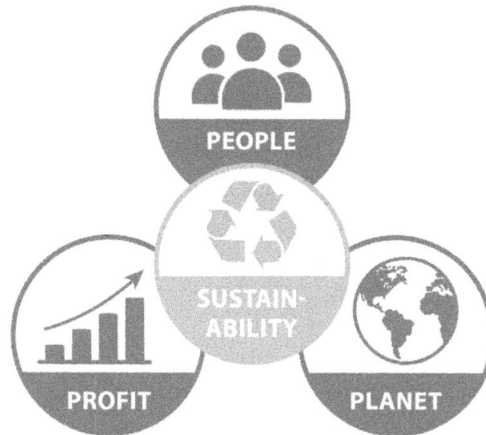

Figure 13.3 – The 3 Ps of sustainability

A sustainable building is focused on improving its overall performance while at the same time, reducing negative impacts on the environment. Technological advances and increasing environmental awareness drive the efforts to make buildings more sustainable – but what is a sustainable building? **The National Institute of Building Sciences** identified six principles of sustainable building design. They are as follows:

- **Protect resources**: Water is a key resource and sustainability methods focus on using water more efficiently by minimizing consumption, recycling, and reusing when possible using IoT sensors. IoT leak detection sensors, smart restrooms, and flow sensors are examples of smart building technologies used for water management.

- **Optimize energy use**: This is a key focus area for sustainability and involves implementing methods to improve energy efficiency and decrease the amount of energy sourced from fossil fuels. Smart building energy management systems use IoT sensors for HVAC, lighting, and other systems to reduce consumption levels.

- **Optimize site potential**: Choosing the right site for the building is impacted by its access to transportation, local ecosystems, and energy sources. A building's orientation can impact the amount of energy required to heat and cool it, and creative landscaping can protect and shield the building while also using IoT sensors to control water use and conserve energy.

- **Enhance indoor environmental quality**: The use of natural lighting, proper ventilation, temperature, and moisture control all impact occupant comfort, productivity, and health. Each of these systems will use IoT smart building solutions.

- **Optimize building space and material use**: This involves using materials and optimizing the use of space and resources by using IoT-driven space planning and management tools.

- **Optimizing operational and maintenance practices**. This is achieved by engineering materials that simplify operations support practices while reducing consumption requirements and using fewer toxic materials. IoT-driven preventative and predictive maintenance and operations practices reduce resource requirements.

Beyond sustainable concepts, sustainable practices encourage the renovation of existing structures over building new ones, and the recycling of infrastructure by reusing bricks, metal components, wood, wall coverings, copper pipes, and other materials. IoT smart building solutions are used to provide analytical information for building operators to help them meet their sustainability requirements.

To be considered a truly sustainable building, sustainable methodology and ideology must be embedded at each stage of the building's life cycle. These stages include planning, design, construction, operations, maintenance, and demolition.

Sustainability data

Industry sustainability certification organizations such as RESET, GRESB, LEED, and BREEAM require data and accurate reporting across a building's operations. Previous data gathering and reporting practices included manual clipboard recordings which were often prone to inaccuracies and stale data. IoT smart building technology provides real-time information not only for actionable insights but also for sustainability reporting.

Carbon reporting measures the emissions coming from buildings from activities, such as burning gas or oil and electricity consumption. IoT sensors and devices are used to monitor, collect, and report the data on this and for other sustainability-related reporting. Sustainability data must be consistent, accessible, auditable, and timely. In many countries, the data must be in a *machine-readable format*.

The combination of IoT sensor devices, cloud and edge computing, and ML and AI solutions make it possible to efficiently gather and process this data. Digital twin technology allows the development of real-world models of buildings' systems to gather insights on how to make these systems more efficient.

ESG and sustainability confusion

Reviewing the definitions of ESG and sustainability provided earlier, it becomes clear that they are closely related. The 3 Ps of sustainability nearly match the ESG pillars. Recently, these terms have been used interchangeably. While ESG was initially used in reference to investments, the term has become a shortcut acronym for corporate sustainability.

Here are the key differences between ESG and sustainability:

- ESG focuses on the building stakeholders, its identity, and the decision-making processes, while sustainability focuses on the relationship between the building and the environment

- ESG is a framework for investors to review a building's performance and risk, whereas sustainability focuses on capital investments, such as installing LED light bulbs or upgrading energy management systems

- ESG is driven by standards developed by investors, lawmakers, and ESG reporting companies who may choose from dozens of different frameworks, whereas sustainability standards are science-based and standardized, such as the process to measure CO_2

- While ESG includes the sustainability pillar, it also contains broader social and corporate governance criteria

A simple way to remember the difference is that ESG focuses on how the world impacts the buildings while sustainability focuses on how the buildings impact the world. In the past, ESG was more relevant to large buildings; however, more recently it's become relevant to all buildings.

Green buildings

While sustainable buildings and green buildings have similar end goals of being better for the planet, there is a difference between the two. Sustainable buildings focus on the environmental, social, and economic impacts while the processes for green buildings are entirely focused on the environment. These include initiatives such as switching to renewable energy sources, reducing energy consumption, and focusing on carbon footprint reduction.

The **US Green Building Council (USGBC)** defines creating a green building as *"the planning, design, construction, and operations of buildings with several central, foremost considerations: energy use, water use, indoor environmental quality, material section and the building's effects on its site."*

Green buildings are typically awarded LEED or WELL certifications, which focus on standards and best practices that help the building become more environmentally friendly. Not every green building initiative is the same and there are different types of green building technologies. They are as follows:

- **Net-zero concepts**: Net-zero or zero-energy buildings are designed to produce and supply their own power. They use renewable energy sources to reduce their carbon footprint – that is, achieve zero carbon emissions when building.

- **HVAC**: Green buildings use HVAC systems that are designed to lessen electricity consumption by using IoT smart building technology and automation solutions.

- **Low-emitting materials**: These green buildings have strategic goals to reduce the use of toxic-emitting materials to protect the environment and improve human health.

- **Cool roofs**: Cool roof technology reflects the heat and sunlight away from a building to regulate the building and room temperatures, which reduces the amount of electricity for HVAC systems to use.

- **Green insulation**: These green buildings use eco-friendly wall fillers for their wall and floor insulation to deliver energy-efficient heating.

- **Solar power**: This is one of the most common sustainable green building technologies used today with the intent of making the building become self-sufficient using a renewable and clean energy source.

- **Smart appliances**: Energy-efficient smart appliances, such as smart refrigerators, dishwashers, and washing machines, can help the building achieve self-sufficiency or meet the net-zero concept.

- **Water efficiency**: As defined earlier, one of the principles of a green building is water conservation. The use of potable and non-potable water for different uses can take advantage of gray water or rainwater for the non-potable water supply.

Sustainable and green buildings have common environmental impact goals to use less energy, produce less waste, and use less water than traditional buildings. Both are focused on reducing or eliminating any negative environmental impact and can be applied to renovating a building or building a new one from scratch. Residential buildings, commercial buildings, industrial buildings, and every other building type are candidates as well.

EU sustainability taxonomy

The European Green Deal has established climate and energy targets for 2030. The **European Union (EU)** has established a common classification system for activities related to sustainability. This taxonomy provides definitions for environmentally sustainable economic activities to companies, investors, and policymakers. EU taxonomy is expected to become a bigger issue to deal with than ESG, as it is a regulation that will impact 50,000 companies and every supplier to these companies.

The EU taxonomy regulation has identified six environmental objectives as follows:

- Climate change mitigation

- Climate change adaptation

- The sustainable use and protection of water and marine resources

- The transition to a circular economy

- Pollution prevention and control

- The protection and restoration of biodiversity and ecosystem

The **Technical Expert Group** (TEG) on sustainable finance has developed recommendations for technical screening criteria related to economic activities in order to reduce climate change and prevent causing significant harm to the EU objectives. In March 2020, the group finalized recommendations for the EU taxonomy and an implementation guide for financial institutions.

They established technical screening criteria for 70 climate change mitigation and 68 adaptation activities, along with criteria for doing no harm to the environmental objectives. Their recommendations also provided a methodology section, Excel tools, feedback processes, and frequently asked questions.

ESG and smart buildings investments

In a sense, it's the perfect storm coming together. Challenging economic conditions, a post-pandemic new-normal hybrid working model, increasing energy prices, and governmental tightening of energy regulations are forcing investments in IoT smart building solutions. New technology is being developed specifically to help building owners and operators meet ESG goals.

Building Management Systems (BMSs) use IoT technology to embed physical objects with sensors to become more energy-efficient and collect massive amounts of data to provide actionable insights. ESG and sustainability have become a high priority with a worldwide consensus that we need to achieve net-zero goals globally by 2050. These drive smart building investments and PwC estimates there will be an 84% increase in ESG-labeled investments by 2026.

Buildings adopt IoT smart building technologies to help achieve their ESG objectives. IoT and BMSs are traditionally separate and usually complex ecosystems. The common denominator is data, automation, and real-time analysis to deliver actionable insights. BMSs unfortunately use a range of different technologies, often have their own proprietary languages, and are siloed from other systems. Multi-site organizations will have different vendors for their BMSs, which adds additional complexity.

This is where IoT and smart building platforms must come together to integrate all the buildings' systems to operate, measure, and report to meet ESG goals. It will most likely not be financially viable to replace these disparate systems all at once. Therefore, they must be integrated with local hardware, software APIs, and a common communications system, along with devices to visualize these results on a common dashboard. Open source software and security should always be considered to protect the physical components and the data. IoT presents an opportunity for cyber-attacks.

IoT and smart building solutions deliver opportunities to manage buildings' services, minimize environmental impacts, and reduce costs. They will also help measure and report ESG metrics. Risks around scalability, data governance, and the ease of integration will need to be anticipated.

Sustainability funding

As we might expect, there is a cost to running an ESG or sustainability program for a building. Fortunately, regardless of your location, there is most likely a government-sponsored program to assist in defraying some or all these costs. We recommend that you review the programs available to you from your local, regional, state, provincial, territory, national, and federal government agencies. Often, utility providers will have rebate or discount programs available as well.

Sustainability funding programs come in many different formats, from money provided in advance to credits given later. Some of the program types we recommend you search locally include the following:

- **Tax credits**: For example, in the United States, there are federal income tax credits and other incentives for energy efficiency along with renewable energy tax credits. IEA.org has 31 member countries and 8 associated countries and they help to locate tax credit programs for sustainability development.

- **Sustainability research grants**: These grants support the research and development of methods, guidance, tools, and programs that benefit the application of sustainability.

- **Sustainability grants**: Most countries offer grant programs for sustainability initiatives:

  - The United States 2022 grant program has over $4 billion available

  - Canada's SDG funding program provides up to $100,000 for 12 months

  - The EU partly funds projects, along with other private sources

  - There is also the **European Fund for Sustainable Development Plus (EFSD+)**

  - Australia's Grant Connect list numerous sustainability grants available and each state and territory have its own grant programs available

- **Green infrastructure funding**: Government agencies such as the United States **Environmental Protection Agency (EPA)** make funds available. The **Small Industries Development Bank of India (SIDBI)** offers a sustainability finance scheme. Invest India is a sustainability framework and funding is made available by the government.

- **Subsidies**: The EU has subsidies managed by national and regional authorities.

- **Private grants and funds**: There are several private philanthropic organizations seeking an independent world that offer grants and funding for sustainable projects. The Rockefeller Brothers Fund, A Glimmer of Hope Foundation, and the African Development Foundation are a few examples. `fundsnetservices.com` provides a list of many of these groups.

Sustainability projects should not be discarded due to lack of funding, as there are many creative alternative ways to raise funds. Potential donors are particularly attracted to sustainability strategies. Shared savings agreements have provided hospitals with energy reductions by adding a third party to finance, design, and implement retrofit projects. With a little research, you may be able to find unique programs in your area.

Global sustainability plan

In 2022, US President Biden issued Executive Order 14057, entitled *The Federal Sustainability Plan*, which established a set of goals to reduce greenhouse gas emissions. These goals outlined here are in line with the updated Paris Agreement goals adopted in September 2022 by the 194 member countries of the **United Nations Framework Convention on Climate Change (UNFCCC)**. It is therefore more than likely that wherever you are reading this book, your country has adopted the same or similar goals.

The importance of understanding these goals is that buildings play a major role in causing greenhouse gas emissions and therefore they will need to play a significant role in reducing these emissions to achieve these goals. IoT and smart building technology will be required throughout buildings to this end.

The sustainability goals are as follows:

- **100% carbon-pollution-free electricity (CFE) by 2030, including 50% on a 24/7 basis**: This will impact the energy purchasing decisions made by building owners and operators.

- **100% zero-emission vehicle (ZEV) acquisitions by 2035, including 100% light-duty acquisitions by 2027**: This will include a building's fleet of service trucks and equipment. It also means that buildings' vehicle charging infrastructure will need to be expanded.

- **Net-zero emissions buildings by 2045, including a 50% reduction by 2032**: To achieve this goal, energy and water efficiency will need to increase while the use of toxic materials and production of waste will need to be reduced or minimized. New markets for recycled products will need to be pursued and owned and leased buildings will need to achieve higher levels of sustainability.

- **Net-zero emissions procurement by 2050**: Programs to buy clean low-carbon materials and to maximize the procurement of sustainable products and services will be required.

- **Net-zero emissions operations by 2050, including a 65% reduction by 2030**: Buildings will need to achieve net-zero emissions by 2045.

- **Climate-resilient infrastructure and operations**: Processes to routinely assess climate vulnerabilities and risks will be required.

The United Nations has identified the building sector as the sector with the most potential to develop solutions that can play a major role in reducing greenhouse gas emissions. It is estimated that the use of energy in buildings accounts for 20% to 40% of all global emissions and therefore IoT smart building solutions that improve energy efficiency will attract positive attention.

Summary

Nearly 200 countries have agreed to sustainability goals to reduce greenhouse gas emissions. Buildings are one of the largest sources of gas emissions and therefore they will be central to achieving sustainability and ESG goals. IoT and smart building technology will be required to monitor, manage, and integrate all buildings' systems to reduce resource consumption and pollutants. ESG and sustainability frameworks help building stakeholders define, measure, and report progress toward their goals.

Sustainable and green buildings both focus on improving the planet. Sustainable buildings focus on the environmental, social, and economic impacts while the processes for green buildings are entirely focused on the environment. To help fund sustainable building projects, most governments around the world offer tax credits, grants, funding, and subsidies.

The next chapter will focus on how a building is a microcosm of a city with many of the same needs, such as managing water, energy, and lighting or delivering emergency services and providing security. With this in mind, it makes sense to build smart cities by beginning with building smart buildings.

14

Smart Buildings Lead to Smart Cities

Resource consumption reduction, operations efficiency gains, environmental improvements, sustainability, and enhancing the quality of life for all stakeholders are just a few of the smart city goals that match the goals of smart buildings. Buildings often operate as small cities within themselves, having to address nearly all the same systems in a complementary manner.

Buildings are a microcosm of cities with similar needs to manage resources, water, energy, lighting, emergency services, security, and other services. Along the same line of reasoning, smart buildings are a microcosm of smart cities and therefore serve as the ideal launching point to grow and develop smart cities.

Smart cities contain complex layered systems and an ecosystem of networks. These network foundations may be built with individual buildings that each use IoT smart building solutions to address building automation, energy optimization, and numerous other smart outcomes. These buildings are the building blocks of scalable smart cities.

In this chapter, we're going to cover the following:

- Define what a smart city is
- Review the benefits of a smart city
- Demonstrate how smart buildings are the foundation of a smart city
- Present a framework for measuring the impact of a smart city
- Explore how city and government buildings are the perfect launching points for building a smart city foundation
- List the five smartest cities in the world and the buildings that helped these cities achieve their ranking
- Understand why many smart city initiatives fail

The elusive smart city

There has been much talk about smart cities over the past decade but much fewer results have been delivered than were promised. Even prior to the pandemic in 2020, smart city initiatives fell short, with single-purpose projects merely centered on reducing city lighting costs, delivering neighborhood Wi-Fi, or providing bike-sharing services. To be fair, there are a few exceptions, and we will discuss a couple of these later in the chapter, but otherwise, many smart city initiatives have tended to fall short of their intended goal.

Google's Sidewalk Labs in Toronto was a billion-dollar smart city grid loaded with sensors and cameras envisioned to revitalize the post-industrial shoreline. While many will claim the project was shut down due to COVID-19, Bennat Berger summarized in his article *Sidewalk Labs' Failure and the Future of Smart Cities* that the real reasons were the residents' fears of data privacy and the inability to make the project financially viable without sacrificing significant parts of the experience.

North Carolina's Union Point Park promised smart microgrids, self-driving vehicles, and smart parking, building high expectations but delivering very little, as highlighted by John Lorinc in his article *Smart City megaprojects get a lot of hype. So why do so many turn out to be expensive disappointments?*. Developers pulled out of the project amid protests and petitions from city residents, who claimed their needs were never heard. The project was greenlighted by city officials with the goals of job and economic creation using the latest whiz-bang technology, but it never addressed the citizen stakeholder needs.

Numerous Asian cities that attempted to build smart cities from the ground up are now technology ghost towns. These technology-led projects looked great on paper yet turned out to be impractical and unrealistic in application, as residents did not find the features useful, and the projects were rushed for the mere sake of implementing technology.

Original Equipment Manufacturers (OEMs) have dazzled cities with trial and pilot projects around smart lighting, smart waste management, Wi-Fi coverage, bike-sharing, and others, only to shut down after the first few blocks are completed due to a lack of city funding to expand these programs. While these projects look great initially, they often do not help politicians get re-elected when residents believe money should have been spent on other city priorities, such as supporting the homeless and reducing crime.

Smart city projects fail for many reasons, but most often, it is due to either a lack of funding, concerns around the invasion of privacy, perceived or real social discrimination, and/or failure to educate citizens on the project's benefits. Many cities are sold on technology first and convinced of other benefits later to try to justify the spending. These projects are not coming from a position of *what's the problem, here's the solution*. Rather, they are often sold on *here's some cool technology, let's find a place to install it*.

City organizations are structured like any company or building organization with responsibility dispersed among multiple leaders. Different departments are narrowly focused on solving each of their own issues, often in a vacuum. Smart city projects are approved by different departments and typically not integrated into a master smart city plan. Like the issues faced by smart buildings, they are often siloed, disparate projects that may not even solve or prevent real problems or enhance the citizens' quality of life.

Smart city defined

Futuristic-looking buildings with unique exteriors, self-driving vehicles, smart traffic lights, air quality monitors, streetlights that light with motion sensors, and other imagined amenities that help a city run to perfection are the images that many have of a smart city. Movies such as *Blade Runner*, *Her*, and *Minority Report* have conceptualized smart cities in different ways, but the one thing they have in common is the use of IoT technology and artificial intelligence.

Figure 14.1 – Futuristic city image courtesy of WallpaperAccess

Let's begin by understanding what a smart city is. Like trying to define smart buildings in the first chapter, there are several definitions for a smart city. The following list provides some definitions from different organizations:

- **IBM**: "One that makes optimal use of all the interconnected information available today to better understand and control its operations and optimize the use of limited resources."

- **TWI**: "A smart city uses information and communication technology (ICT) to improve operational efficiency, share information with the public and provide a better quality of government service and citizen welfare."

- **Cisco**: "A smart city uses digital technology to connect, protect, and enhance the lives of citizens. IoT sensors, video cameras, social media, and other inputs act as a nervous system, providing the city operator and citizens with constant feedback so they can make informed decisions."

- **Earth.org**: "A smart city is a concept that sees the adoption of data-sharing smart technologies including the Internet of Things (IOT) and information communication technologies (ICTs) to improve energy efficiency, minimize greenhouse gas emissions, and improve quality of life of a city's citizens."

- **United Nations Economic Commission for Europe (UNECE)** and the **International Telecommunication Union (ITU)** jointly defined smart sustainable cities as: "An innovative city that uses information communication technologies (ICTs) and other means to improve quality of life, efficiency of urban operation and services, and competitiveness, while ensuring that it meets the needs of present and future generations with respect to economic, social, environmental as well as cultural aspects."

We've combined the common elements from these definitions to create our definition: *The Internet of Things (IoT) and Information and Communication Technology (ICT) are interconnected to collect data/information used to improve city operations, reduce consumption, improve the environment, and enhance citizens' quality of life.*

With many different yet similar definitions of smart buildings, it became necessary to adopt a common reference around the world to conceptualize and benchmark smart cities' impact. Dr. Rudolph Giffinger developed six common dimensions of a smart city to evaluate their impact:

- **Smart people**: People are defined by their education or qualification levels, their social interactions, and their openness to creative innovations from outside the city. This dimension involves a more inclusive society to improve access to education, knowledge management, and social capital.

- **Smart mobility**: Mobility includes local and international accessibility, sustainable transportation systems, and the availability of ICT to meet the needs of all users (residents, workers, tourists, citizens, and so on).

- **Smart governance**: This comprises citizen services, political participation, and how the administration functions. It focuses on making public actions open and transparent. Stakeholders are encouraged to engage in the decision-making process.

- **Smart environment**: This dimension includes locating the building close to attractive natural conditions, natural and heritage resource management, pollution management, and environmental protection practices. The focus is on resource sustainability management (food, water, energy, air, and so on) along with the production of renewable green energy.

- **Smart economy**: A smart economy looks at economic competitiveness factors, such as entrepreneurship, innovation, labor market flexibility, trademarks, and the integration of other markets. The circular economy and sustainable entrepreneurship are part of this.

- **Smart living**: This dimension addresses the improvement of citizens' quality of life, which includes culture, housing, tourism, safety, and health.

There are several other soft factors related to these common dimensions, which include education, industry, participation, and technical infrastructure. These dimensions are complementary and smart city projects may contain one or more.

The benefits of a smart city

The benefits of a smart city are like the benefits of a smart building and include the following:

- Lowering operating costs for city-owned buildings
- Decreasing greenhouse gas emissions
- Increasing tax revenue
- Attracting/retaining new and targeted businesses
- Creating new digital jobs
- Improved safety
- Increased city resilience
- Improvement in city management and city infrastructure
- Increased city and citizen engagement
- Enhanced city reputation

The benefits of a smart city go beyond the traditional financial metrics, with outcomes that enhance the quality of life for its citizens. Cities are typically cost centers, so they rely on tax revenue and offset costs by using IoT and smart technology. Smart buildings help reduce costs, lower emissions, and attract new revenue opportunities.

Smart buildings create the smart city

A building is a microcosm of a city in many ways. Like connecting systems within a building and using the data collected to deliver actionable insights, connecting buildings across a city takes this a step further to identify inefficient use patterns across several buildings. Cities already exist and buildings already exist; therefore, they will just need to be upgraded with IoT technologies to collect data to make insightful adjustments.

Smart building technology deployed across the city can help streamline operations and improve the overall infrastructure to deliver the smart city. Smart cities use the same advanced digital technologies and analytics as smart buildings, with a focus on similar outcome areas, such as the following:

- **Energy**: Cities are establishing goals to provide more efficient use of the power grid, something that is already occurring in many buildings. Buildings are a large part of a city's energy use, and cities are responsible for securing and providing energy supplies to these buildings.

Individual buildings should analyze energy use data, share this information with the city, and adjust accordingly to avoid peak-hour energy charges, which can cost up to 30 percent more than non-peak hour. Cities are responsible for the energy effects on the environment, and the use of IoT monitoring and smart city energy management systems can have a substantial impact.

- **Environment**: Air and water IoT sensors deployed across the city's buildings and the grid store data on the quality and chemical composition and alert field personnel of issues. Like buildings, cities are focused on reducing carbon emissions and use IoT-driven smart air quality monitoring solutions to measure, monitor, and manage their impact. A better quality of life can be achieved when the city delivers clean air and water, which can improve the health of citizens.

- **Data**: Smart buildings are built on the foundation of implementing IoT devices to monitor, measure, and report data from physical systems. Integrating all the building's systems generates enormous amounts of data used to improve efficiency and reduce consumption. On a much larger scale, cities are integrating city systems the same way buildings have, and the outcome is a massive amount of data that is used to improve the lives of its citizens.

  Cities will build much of their data by repurposing existing data and digital documents from the built environment. This data is already captured by cities in various formats by their internal organizations, including planning departments, tax departments, building departments, permits and licensing departments, engineering departments, land use departments, and even the post office. When all this data is connected and analyzed, smart city solutions can be developed.

- **Digital services**: Cities and buildings need to deploy seamless electronic and internet services for navigation, wayfinding, asset tracking, and people's locations. Digital interaction allows city officials and building managers to electronically interact for convenience. First responders can save lives and reduce asset and building damages using digital twin data.

- **Healthcare**: The capabilities of hospitals and other medical facilities across the city are expanded to deliver remote access to healthcare, video consultations, medical records, and medical robots. Better patient outcomes can be achieved with smart wearable devices to monitor vital signs. Communicating real-time data directly to specialists and other medical professionals reduces critical wait time and reduces expenses. City-wide analytics will address big issues and trends, such as COVID-19, and deliver actionable outcomes. Smart building/city solutions also deliver methods to alert citizens of health and safety risks.

- **Location services**: Many manufacturing facilities, commercial buildings, shopping centers, airports, entertainment centers, and stadiums use location services for navigation, wayfinding, and asset location. Smart cities are now using these smart building location services on a much larger scale to address smart city infrastructure needs and deliver city-wide services.

  On the other side, mini cities exist whereby the smart building infrastructure acts the same as the smart city infrastructure. Examples include large shopping malls, resorts, cruise ships, and, more recently, mixed-use residential and commercial spaces. These locations require large location service systems like those of massive urban areas.

- **Microgrids**: Smart buildings connected and talking to each other create a microgrid that can help stabilize the energy grid and better manage fluctuations and demand. These microgrids monitor energy supply and demand and manage the distribution of power. Interconnected with smart buildings, these grids learn the building's energy needs, match the current and predicted weather conditions, and automatically adjust. They may automatically limit consumption when the energy grid is over capacity. The efficiency gains contribute to meeting sustainability and emission goals.

- **Safety and security**: Safety and security are critical to both buildings and cities. City-wide public safety efforts are now commonly enforced in buildings. Buildings are now required to provide enough connectivity to support first responders. They are also required to provide the exact location of first responders and others in the building, regardless of what floor they may be on. Cities are now including accurate locations in their smart city programs similar to how buildings are already delivering these accurate locations.

  Building-mounted IoT video cameras and video analytics are used to monitor and manage crowds and traffic and observe suspicious objects or people. CCTV cameras mounted on buildings detect situations, crimes, accidents, and potential threats. Face recognition helps police locate and track perpetrators in crowded areas.

  Acoustic gunshot detection systems (**AGDSs**) improve public safety, and mobile phone GPS systems alert building occupants and citizens when they might be near emergency scenarios or dangerous areas. Digital twin technology delivers real-time and stored building information, asset management, and people's location information.

- **Transportation**: Buildings are often part of the city's transportation plan, which may require the building to have its own transportation management program. Flow into and out of the building's grounds, parking lots, and delivery docks, along with public transportation and ride-sharing, are all examples of building plans that impact the city.

  IoT sensors, cameras, and real-time analytics are IoT-driven smart building and city solutions that have a major impact on the city's livability and economy. Traffic data collection and analysis systems required to manage traffic patterns across cities are already being used at a smaller scale in sports venues. These venues use the quality of cell signals in a certain area to determine whether they should set up a service kiosk there. Cities use the same technology to determine overcrowded areas at certain times so that action may be taken to widen sidewalks, increase public transportation, or dispatch crowd control personnel.

- **Waste management**: Waste containers across buildings throughout the city contain an IoT sensor to track and report waste levels. Waste is collected when the sensor reaches a targeted level, thereby eliminating the pickup of empty or partially filled containers.

- **Utilities**: Water is one of the most precious resources, and therefore, city-wide water conservation and management across all buildings is required. Smart sensors identify leaks to reduce the time to repair and reduce the amount of wasted water. Smart meters provide citizens with water, gas, and electricity data to ensure they are billed accurately and to encourage conservation.

Buildings and cities will have the same goals in each of the areas listed here, and while their scope may be slightly different, they will be very complementary. Buildings are an extension of the city's environmental furniture, which can be used to host IoT sensors, devices, cameras, and other smart technology. The next section focuses on a framework used around the world for smart city self-assessment, strategic planning, and implementation.

Framework for measuring smart city impact

A successful smart city initiative may be measured based on several different potential outcomes. Similar to a holistic smart building, a holistic smart city will implement smart programs that generate positive outcomes across many different performance factors, and they need a framework to measure and report these outcomes.

A **holistic key performance indicator** (H-KPI) framework has been developed by the US Department of Commerce's **National Institute of Standards and Technology (NIST)**. This framework is used to measure the return on investment and community impact of certain technology or projects across an entire *smart city ecosystem*.

The framework involves assessing data across three main levels of analysis:

- **Technologies layer**: This level includes the city's sensors and actuators, networks, data systems, and computational hardware and software systems

- **Infrastructure services layer**: This includes the communication, transportation, energy, water, and building sectors, as well as related services, including emergency response, law enforcement, waste management, education, and city/community services

- **Community benefits layer**: This layer includes applications that benefit people and businesses and provide equitable access, including for personal safety and security, business and jobs growth, healthcare, environmental quality, and other quality-of-life factors, such as arts and entertainment

Once these assessments across the layers are completed, they will be measured against the framework's five core metrics. These metrics are as follows:

- The alignment of KPIs with the city's priorities across the city's districts and neighborhoods

- Alignment with the city's investment priorities

- Investment efficiency analysis

- Information flow density

- Quality of infrastructure services and community benefits

Cities will want to compare their smart programs with other cities and therefore need a common methodology and framework to do this. One recommended method is to identify and measure KPIs determined for smart cities using KPIs linked directly to the United Nations' 17 **Sustainability**

Development Goals (SDGs) for 2030. The SDGs were adopted by the United Nations as a blueprint for peace and prosperity today and in the future, as a call to action in a global partnership. These KPIs are as follows:

Figure 14.2 – United Nations' 17 SDGs

The KPIs are divided into core KPIs and advanced KPIs. All cities that want to become smart must reach at least the core indicator levels, and cities that wish to achieve a higher level must reach the set of advanced indicators. These core and advanced KPIs are divided into three dimensions:

- **Economy**: The smart city economy is measured through various indicators that represent the smart city level. ICT is vital in collecting data and acquiring it quickly. This data must be trusted by all.

- **Environment**: KPIs focusing on the environment are vital for planning a smart city and delivering the SDGs' targets through specified actions. Capturing and inputting results into well-designed and structured technological networks is required to provide reliable and rapid results.

- **Society and culture**: To achieve a well-established smart city, programs focused on society and culture performance must be implemented.

Figure 14.3 lists some of the possible smart city core KPIs, each tied to the economy, environment, or society and culture performance area:

Economy-Related KPIs	Environmental KPIs	Society and Culture KPIs
Core smart city economic performance KPIs:	**Core smart city environmental performance KPIs:**	**Core smart city society and cultural performance KPIs:**
• Household internet access	• Air pollution	• Cultural expenditure
• Fixed and wireless broadband subscriptions	• Greenhouse gas (GHG) emissions	• Informal settlements
• Wireless broadband coverage	• Electromagnetic fields (EMF) exposure	• Gender income equality
• Smart water meters	• Green areas	• Gini coefficient
• Smart electricity meters	• Renewable energy consumption	• Disaster-related economic losses
• Dynamic public transit information	• Electricity consumption	• Police service
• Traffic monitoring	• Residential thermal energy consumption	• Fire service
• R&D (Research and Development) expenditure	• Public building energy consumption	• Violent crime rate
• Patents		
• Public transport network		
• Bicycle network		

Figure 14.3 – Core smart city KPIs

Figure 14.4 lists some of the possible advanced smart city KPIs, each tied to the economy, environment, or society and culture performance area:

Economy-Related KPIs	Environmental KPIs	Society and Culture KPIs
Advanced smart city economic KPIs:	**Advanced smart city environmental KPIs:**	**Advanced smart city society and culture**
• Public Wi-Fi	• Noise exposure	• Electronic health records
• Electricity supply ICT monitoring	• Green area accessibility	• In-patient hospital beds
• Open data	• Protected natural areas	• Health insurance/public health coverage
• E-government	• Recreational facilities	• Cultural infrastructure
• Public sector e-procurement	• Residential thermal energy consumption	• Housing expenditure
• Transportation mode share		• Child care availability
• Travel time index		• Resilience plans
• Shared bicycles and vehicles		• At-risk population
• Low carbon emission passenger vehicles		• Emergency service response time
• Public building sustainability		• Traffic fatalities
• Urban development and spatial planning		• Local food production

Figure 14.4 – Advanced smart city KPIs

While the NIST framework suggests the preceding KPIs, I have found that there are additional ones, such as business retention, business attraction, technology investments, and innovation hub development, that can be added based on the city's priorities and goals.

IoT helps buildings give back to the grid

Integrating buildings and the electricity grid is a fundamental method for increasing energy efficiency across smart cities. Smart sensing, metering, monitoring, and management are helping building owners to locate efficiency opportunities and to increase the grid operator's situational awareness. The energy savings by using **demand-responsive devices** are augmented with potential savings from not having to build new energy generation and transmission infrastructure.

A *transactive* approach to energy allows millions of meters, sensors, and smart appliances to seamlessly communicate and coordinate with energy loads and distributed generation sources. Decisions for allocation and use are made based on the value, which may use non-energy criteria, such as the power's *greenness*, comfort, and asset valuation.

To build integrated building-grid ecosystems, cities and buildings should do the following:

- **Use the built environment**: Existing buildings can act as low-cost storage options to balance demand peaks and valleys seamlessly.

- **Open inaccessible markets**: Building owners, operators, and occupants should be given access to currently inaccessible energy markets. By doing this, new cash flow opportunities can be achieved from energy savings, energy trading, and efficiency gains, along with the valuable insights gained from the energy use data.

- **Open new markets**: Opening new markets to new participants can lead to new economic and technological development.

Building-to-grid (**B2G**) or **grid-integrated buildings** create value across key stakeholders, which include building owners, operators, utilities, grid operators, and building occupants. Optimizing buildings and integrating them into the grid is good for business and the environment.

Net-zero buildings

Net-zero buildings (**NZBs**) can provide energy storage, flatten peak loads, and provide clean energy back to the grid to make the grid more sustainable and resilient. While many NZBs use solar energy to provide an average of about half the building's required energy, solar energy is only available instantly, and typically, cannot be stored for use later. Any excess solar energy that is generated is sent to the grid in trade for fossil fuel-generated grid energy required during non-solar generating times. This back-and-forth energy trading makes the building *net zero*.

NZBs contribute to the grid in the following ways:

- **Demand response**: To reduce consumption during peak load periods, building operators reschedule certain operations to off-peak times. Another method is to use battery power during peak times and recharge the battery during off-peak periods. This process keeps fossil fuel-powered energy from being used.

- **Ancillary service markets**: Buildings may get paid to participate in frequency regulations markets, capacity markets, and other grid-reliability and stability support services programs. These programs are designed to consume or deliver power with the intent of retaining it for use later.

- **Energy storage**: As mentioned earlier, solar energy that cannot be used right away or stored is traded for other energy that can be stored in the building and used later. This also facilitates a building's participation in ancillary service markets and demand response programs.

- **Microgrids**: Several buildings are connected to share their energy resources. This improves reliability through resource sharing and redundancy. These grids also support self-generation when the grid is offline.

NZBs have enormous potential in widespread sustainable grid deployments. They should focus on installing IoT smart meters along with energy storage and solar power to participate in the smart city energy grid.

Government buildings are smart city starters

Nearly every city has numerous buildings that are owned by the city and therefore present opportunities to build the foundation for a smart city. These buildings include courthouses, jails, city halls, libraries, police stations, tax offices, farmer's markets, and city operations. Cities will often also contain provincial, state, territory, and federal/national buildings.

A government building operates the same as a private building, and therefore, all the efficiency and cost-cutting initiatives we discussed for buildings apply to these buildings as well. Cities wishing to implement smart city initiatives should consider their own buildings as launching points. These government-owned and managed buildings will require all the same IoT devices and smart technologies that have been discussed throughout this book.

The US **General Services Administration (GSA)** is the US government's largest civilian landlord. They have built a smart building technology framework to increase tenant satisfaction, lower operating costs, improve cybersecurity, and allow operations and facility management to manage the buildings using the building's data. To achieve this, GSA is deploying technologies such as the following:

- **Advanced metering**: Using IoT meters, they record and communicate energy consumption across all their buildings. This provides insight into consumption behavior and customer billing, and allows for system monitoring.

- **GSAlink**: This IoT-driven **fault detection and diagnostics (FDD)** system investigates trends, patterns, and faults in the operational data to ensure proper building system operations and improve efficiency.

- **National Digital Signage** (NDS): Tenants are informed about facility news, weather, and other critical real-time emergencies via digital kiosks deployed throughout the building.

- **National Computerized Maintenance Management System** (NCMMS): Work orders, preventive maintenance, building logistics, and equipment management are a few of the many systems that have been integrated and are now managed through a single management system.

- **Deep energy retrofits**: These government building retrofits are designed to reduce energy use and costs by 40 percent per building. These are integrative whole-system approaches, rather than trying to bridge individual technologies in isolation.

- **Energy and water management program**: This program uses IoT monitoring sensors and is focused on promoting optimal energy use to reduce federal utility costs.

GSA, along with other government-owned building operators, is implementing smart building executive orders to ensure their smart building programs align with their strategic sustainability goals. These orders include the following:

- **Modernization**: Facility management, operations, and maintenance practices are being modernized to provide cost reductions to reduce the taxpayer's burden.

- **Environmental impact**: These orders drive building initiatives that limit the impact on the environment and support energy conservation.

- **Technology**: Orders are issued to use the latest available technology to measure, monitor, and manage the building's elements that impact tenant comfort and building occupant satisfaction.

- **Data**: This order requires that building data collected using IoT devices is available to government staff and contractors to be used to make informed building management decisions.

- **Interoperability**: This order requires the use of open protocol systems to promote interoperability between devices and systems delivered on a facility-wide tool.

- **Cybersecurity practices**: Government buildings are ordered to implement and maintain best practices for cybersecurity for all national, regional, and local offices. These must be consistent with the government's internet protocol network-based system and all downstream devices.

- **Risk management:** All government smart building initiatives must protect the government and must use principles developed specifically for **cyber supply chain risk management (C-SCRM)**.

These orders should drive consistency to enable opportunities to share resources, propagate best practices, and advance appropriate practices for the delivery of innovative new smart building systems.

World's top smart cities 2022

As mentioned earlier, there are some cities that have been very successful with their smart city initiatives. The **Smart City Index** is an annual report developed by the **Institute for Management Development** and the **Singapore University for Technology and Design (SUTD)**, which ranks cities using technological and economic data, along with a perception rating from the citizens on how *smart* they perceive their city to be.

The top-ranked cities and smart buildings in 2022 were the following:

- **Singapore**: This city-state launched its smart city initiative in 2014 in both the public and private sectors. Citywide smart programs include a digital health system that uses IoT wearable devices and video consultations, a contactless payment system for public transportation, and a portion of the city designated as vehicle-free.

  Smart buildings include Capital Tower, a 52-story building and winner of the Green Mark Platinum Award for design, energy, and water efficiency, and Singapore's Changi Airport, which installed an airside Wi-Fi apron to connect devices for asset tracking, access digital service manuals in real time, and provide airlines with real-time flight information.

- **Helsinki, Finland**: This city has set aggressive carbon-neutral goals for 2035, 15 years ahead of the Paris Agreement. Heating accounts for over 50 percent of the city's emissions, so they have focused on implementing energy-efficiency programs to reduce emissions by 8 percent. City-owned buildings are focused on using renewable energy, and data is published as open data.

  Helsinki Marketing worked with Finlandia Hall as a test venue, to focus on event energy consumption and water consumption, waste sorting, and food and drink sustainability indicators for the amount of organic ingredients, responsibly sourced food, and water.

- **Zurich, Switzerland**: Launched in 2015, several buildings interconnected electricity, heating, and cooling systems that are automatically controlled through an intelligent management system to reduce CO_2 emissions. They have built a green district of residential buildings using renewable energy sources, such as roof-mounted photovoltaic panels connected to a smart grid, where extra energy is used for charging electric vehicles.

  DFAB HOUSE built the **Next Evolution in Sustainable Building Technologies (NEST)** building of Empa and Eawag outside Zurich. NEST is a building platform where IoT and smart building technologies are tested and demonstrated. Innovative technology can be plugged and played together.

- **Oslo, Norway**: The focus here was on retrofitting existing buildings with green energy systems and circular waste management. Projects also included zero-emission construction sites. The M:6 building is a co-working open-plan environment along with traditional offices designed to incorporate new technologies. The center is using IoT technology to improve and manage resource consumption along with predictive building operations and maintenance. IoT door sensors, electricity meters, and water sensors provide alarms for non-typical situations. A common platform is used to integrate different technologies.

- **Amsterdam, The Netherlands**: The smart city program started in 2009 and now involves 170 different city operations. Highlights include solar-powered bus stops and thousands of businesses and homes modifying their roofs with energy-efficient roofing insulation, as well as adding smart meters, LED lighting, and IoT motion sensors to dim lighting.

  The Deloitte Edge building is considered by most experts to be the smartest building in the world. It has IoT smart capacity meeting rooms, working tables, and parking. Sensors measure light level, temperature, and humidity, and smart restrooms are connected directly to the cleaning team. Solar panels and smart blinds manage the 15 floors to optimize energy consumption. All systems are connected to a single platform to collect, share, and store data.

Other notable smart buildings around the world helping to make their cities smarter include the following:

- Tottenham Hotspur Stadium – London, UK

- Sheraton – Los Angeles, US

- The Crystal – London, UK

- Apple Park – Cupertino, US

- Bahrain World Trade Center – Manama, Bahrain

- Allianz Arena – Munich, Germany

These smart cities and smart buildings are being driven in large by the IoT smart building technologies that have been discussed throughout this book, such as using IoT sensors and devices to monitor and manage the building, and by connecting and integrating these buildings within the city.

Summary

Cities continue to evolve and grow, and they have an enormous impact on the people who live and work in them. Buildings and cities have similar goals around resource consumption reduction, operations efficiency gains, environmental improvements, sustainability, and enhancing the quality of life for all stakeholders. Since buildings and cities are working toward the same outcomes, using the same IoT smart technology will have positive and complementary impacts for both.

A building is a microcosm of a city, with many of the same needs, such as managing water, energy, and lighting to delivering emergency services and providing security. With this in mind, it makes sense to consider smart buildings as the building blocks of scalable smart cities.

The next chapter will focus on building smart from the start using the latest smart building technology in the design and construction phases. We'll explore the impact smart buildings are having on the commercial real estate industry. To wrap things up, we'll look at the trends and continued smart building evolution to determine what comes after smart buildings.

15

Smart Buildings on the Bleeding Edge

Smart buildings have become a bleeding-edge technology integration movement in the built environment over the past few years. The 2020 pandemic and the subsequent return to the office saw the need to release many new IoT smart building solutions before they were thoroughly tested and, therefore, came with a degree of risk often associated with bleeding-edge technology. Early adopters weighed the risks and opted to move forward to solve immediate needs around indoor air quality, building access, occupancy levels, and social distancing requirements.

Based on the growing development of IoT sensors, edge/cloud computing, and data management, smart buildings integrate data sources, inputs, and user types with actionable information to create more efficient, effective, and engaging spaces. But what's next? This chapter will explore how the smart building impact is calculated, the evolution of smart buildings, the use of smart construction techniques, the importance of intelligent learning, and where we see smart buildings heading in the future. A final thought will be given to explain the need to align on a data tagging standard to facilitate the future of smart buildings.

In this chapter, we're going to do the following:

- Understand smart buildings' impact on **net operating income** (**NOI**)
- Review the different real estate categories with a focus on how smart buildings are digitalizing the commercial real estate market
- Explore how to build a smart building by using the latest smart construction applications and processes
- Detail the differences between **artificial intelligence** (**AI**), **machine learning** (**ML**), and deep learning and understand how they are making buildings smarter
- Evolve smart buildings into a unified building

- Define how IoT will drive the future of buildings and explore a framework developed to achieve the building's long-term vision

- List a few examples of what we believe the next IoT-driven solutions will solve and their impact on the smart building industry

- Understand the need to align on a data tagging/naming convention for the future of smart buildings

Smart buildings' impact on NOI

NOI is used to determine the potential profitability of real estate investments and is one of the most important **key performance indicators** (**KPIs**) used by building stakeholders. It is typically determined by subtracting all the reasonably required operating expenses from the building's revenue. This before-tax number is included in the property's income and cash flow statements but excludes loan interest payments, capital expenditure, depreciation, and amortization.

The building's revenue includes rental income and all other amenity services' revenue, such as parking, storage, cafeteria sales, vending machines, and laundry. Expenses include the cost of maintaining and running the building, insurance premiums, utilities, legal fees, property taxes, janitorial fees, and repair costs. Capital expenses for large-item purchases, such as a new HVAC system, are not included in the NOI calculation.

Net Operating Income

NOI
Formula = Operating Revenue – Operating Expense

NOI
Formula = Operating Revenue – COGS – SG&A

Figure 15.1 – Formula for NOI

NOI is the same as **earnings before interest and taxes** (**EBIT**), which is a metric used in nearly all other industries. It helps the property owner to determine whether renting a property is worth the expense of maintaining and owning it. The property owner(s) and investor(s) use the capitalization rate to calculate the building's value, which then allows them to compare it to other buildings they may be thinking about buying or selling. The NOI formula is operating revenues minus the **cost of goods sold** (**COGS**) plus **selling, general**, and **administrative** (**SG&A**) expenses.

We have stated throughout this book that successful smart building initiatives address stakeholder pain points. For building owners, NOI calculates potential profitability for their investment and is considered one of their pain points. Smart building projects are often approved based on their potential impact on NOI, and IoT-generated data is used to measure, calculate, and demonstrate the impact on NOI. Smart building initiatives that solve issues, increase revenue, and reduce costs will typically be approved.

In *Chapter 1, An Introduction to IoT and Smart Buildings*, we listed many different types of buildings, from apartments to operations buildings. Here, we are going to focus on some of the large classifications sometimes used to segment real estate owner types into groups. Real estate is defined as property, buildings, land, and other physical structures. Regardless of the owner and building types, each will use NOI to measure their building's return on investment, and each will benefit from implementing IoT smart building technologies. These groups include the following:

- **Owner-operated buildings**: These buildings are typically owned by a company for the use of their own operations. Large corporations may have many buildings used for offices, labs, manufacturing, warehousing, and other activities. Microsoft, for example, has hundreds of buildings around the world, and essentially has its own internal real estate department.

- **Government buildings**: In *Chapter 14, Smart Buildings Lead to Smart Cities*, we discussed how municipal and government buildings are the perfect launching points for building a smart city using IoT smart building technologies.

- **Real estate developers**: Real estate development is the process of making improvements and construction or renovations to raw land and existing buildings to make a profit. Developers may require their construction companies to use real-time job site condition monitoring for safety, quality, and environmental reasons.

- **Commercial real estate**: Commercial real estate includes offices, hotels, malls, apartments, and universities.

- **Industrial real estate**: Industrial real estate includes buildings used for the manufacturing, research, distribution, and storage of goods.

- **Residential real estate**: This group focuses on the construction, purchase, and reselling of single-family homes, vacation homes, and condominiums.

- **Land real estate**: This includes undeveloped land that can be used for development, farming, ranching, and subdivision development.

While each of these groups may focus on different segments, they all can benefit from smart IoT-driven construction technology. Real estate technology is often referred to as **PropTech**, short for **property technology**.

Smart buildings are digitalizing the commercial real estate market

Smart buildings are changing the commercial real estate industry. After years of slow adoption of technology, commercial building owners and operators post-pandemic are realizing the improved outcomes and returns they can achieve by embracing IoT smart building technologies. They can better manage their properties, reduce environmental impact, improve operational efficiencies, and be more competitive. These technologies reduce costs, expand revenue streams, and increase the property's value long-term.

IoT smart building technology is transforming commercial buildings by doing the following:

- Driving sustainability and carbon neutrality with user-centered building designs that also improve occupant well-being
- Creating the opportunity for buildings to perform self-diagnosis, self-correction, and eventually, self-healing
- Delivering highly connected and autonomous systems to improve efficiency and occupant-focused operations
- Merging **information technology (IT)** and **operational technology (OT)** systems to create one system
- Delivering convenience and control to all stakeholders
- Opening new multi-functional spaces and creating real-time flexible space management capabilities
- Changing the way buildings are built with 3D printing, prefabricated, and modular building techniques
- Enabling IoT sensors to collect massive amounts of data for analysis and intelligent reporting

Many businesses have implemented hybrid office and work-from-home approaches, which necessitate integrating home-to-office technology solutions. Access to data enables workers to be more engaged and productive. Commercial buildings must improve connectivity to avoid occupancy rate reductions.

Building smart from the start

Smart construction refers to the use of applications and processes that improve the construction process and the management of projects. Cloud-based technology solutions, collaboration tools, and IoT solutions are used during the construction phase. Digital technologies and industrialized manufacturing techniques come together to lower construction costs, reduce time, improve sustainability, and improve productivity. Technology advances are changing almost every aspect of the building construction industry.

Chapter 8, Digital Twins – a Virtual Representation, highlighted how digital twins and **building information modeling (BIM)** are used in the design phase to replicate the building properties digitally before construction to allow for better design decisions. BIM allows for the virtual tracking of assets and can deliver 3D renditions for contractors and workers to view precise details of the building. In addition to the visual representation, 3D modeling creates numerous layers of metadata, which are used for workflow collaboration, predictive maintenance, and operations.

Figure 15.2 – Digital mobile construction drawings

Many other IoT-driven smart applications exist for the building's construction phase and include such items as the following:

- **IoT cameras and drones**: IoT cameras placed around the construction site are coupled with drones to monitor site conditions, assess safety, and detect problems during the construction phase. Job site cameras monitor construction progress and protect against asset loss. With the use of drones, potentially dangerous project sites may be viewed from a safe distance. Site inspections performed with drones help managers be more efficient.

- **IoT cement/concrete**: *Programmable cement* and *self-healing concrete* have the capability to fix cracks on their own and are becoming commonplace in building construction. Embedded IoT sensors are used to monitor the performance of concrete, including temperature, humidity, and maturity. The IoT sensors communicate the performance data to smartphones and computers wirelessly so that real-time decisions can be made during construction. Knowing the concrete maturity allows for better scheduling and cycling of delivery, work processes, and labor.

  Mixed concrete trucks and delivery companies are using IoT technology to track the dispatched trucks to the building site and to communicate real-time data regarding the concrete mix before delivery. IoT sensors include a thermometer and accelerometer to accurately calculate slump and concrete volume.

- **Robotics**: Autonomous and robotic machinery is used in construction projects with heavy equipment, such as autonomous tracker loaders, land movers, and backhoes. These vehicles use a combination of LIDAR IoT sensors, **Global Positioning System (GPS)**, and **inertial measurement units (IMUs)** technology.

  3D models are created to guide the machinery through the construction site, or workers may remotely control the vehicles via a smartphone or tablet. IoT sensors are placed around the equipment for autonomous self-driving applications. High-resolution cameras collect data used for interactive applications.

- **Sustainability**: Innovations for construction lead to more sustainable designs and construction. Examples include wind and solar power, the use of biodegradable materials, and green insulation materials. One example is the use of IoT sensors embedded in glass windows to alter the amount of radiation that is reflected to reduce energy consumption.

- **Waste and chemical management**: Waste management practices can also impact the carbon footprint on construction sites. IoT sensors measure trash levels to efficiently conduct waste disposal to create space on the site and prevent possible penalties. IoT sensors and devices are used to detect the release of chemicals that are used during the construction phase to ensure they are meeting environmental requirements.

- **3D printing**: 3D printers can print commonly used products using recycled materials to reduce the footprint and increase sustainability. 3D-printed concrete blocks are becoming common. Material sourcing is made faster and easier by printing certain materials onsite and waste is reduced. Prefabrication applications allow for sections to be built offsite and transported to the job site to be used immediately.

Figure 15.3 – 3D printing

- **Thermal bridging**: Thermal bridging is an unintentional path that heat uses to escape from a building, usually bypassing the insulation layer. A simple example is a vent that passes through the roof. While the vent is passing air as expected, the metal of the vent itself is unintentionally transferring heat and reducing energy efficiency. IoT sensors capture loss data and can be used in the digital design phase to identify and correct potential thermal bridges.

- **Connected hardhats**: Smart hardhats use IoT sensors to safeguard sleepy construction workers and can even trace COVID-19 virus exposure. These smart hardhats improve workers' safety and efficiency and use AI to recognize patterns in movements to make continuous improvements.

Smart hardhats can track construction workers' proximity to each other, hazardous materials, and dangerous machinery to improve safety. Microphones, cameras, and sensors create real-time communication capabilities with subject matter experts and remote command centers.

- **Wearables and smart clothing**: Wearable IoT sensors may be worn anywhere on the body and provide additional real-time information through connectivity. Sensors measure environmental elements, location, voltage, and biometrics to track movement and safety. Heads-up displays on smart glasses can be connected to **augmented reality (AR)**, **mixed reality (MR)**, and **virtual reality (VR)** technology to deliver digital representations and renderings.

  Other wearables include smart work boots, gloves, vests, helmets, and watches to monitor fatigue and other health conditions. Smart clothing can detect slips and falls and notify managers immediately or contact emergency responders if needed. They can monitor restricted zones and set off alarms. Smart clothing also monitors workers' vital signs, such as heart rate, temperature, and breathing levels.

- **Exoskeletons**: Exosuits or exoskeletons are metal frameworks fitted with motors that are worn by construction workers to assist them in moving items. Active, unlike passive, exosuits use motors, IoT actuators, and batteries to assist the wearer. Users are less prone to injuries and overexertion. Different versions exist, from a mounted arm, extended arm, supported limb, overhead assist, and whole-body suits to crouching and standing support suits.

Figure 15.4 – Exoskeleton and robotic vest (sources: Clker and Scottish Construction Now)

- **Site sensors**: Construction sites use a variety of IoT sensors to monitor temperatures, noise levels, dust particles, and other environmental elements for safety and comfort concerns. This data is used to support compliance with local, state, and federal regulations.

- **Smart materials**: Many materials used in the construction of a building are now capable of being smart. One very popular material used in smart construction is smart glass. Smart glass contains IoT sensors that allow the glass to switch between transparent and translucent. The switch can happen automatically or through smartphone commands. These smart windows help with thermal management, energy efficiency, and privacy management, by controlling how voltages, light, and heat are applied to the glass.

- **Predictive analytics**: Data from material suppliers, subcontractors, and design plans are continually examined to determine any risk factors that may need to be addressed. Data from past construction projects is added to identify which portions of the project are most at risk.

- **Virtual and augmented reality**: VR and AR are used to train construction workers in a safe and predictable environment. Workers can access guidelines and instructions and physically perform practice simulations on what they just learned with replicated real-life situations. Emergency scenarios are developed, such as natural disasters and equipment malfunctions, to roleplay what should be done. Building plans can be viewed using AR to envision the building's outcome. 3D models are used to plan the design.

- **Structural monitoring**: Structural monitoring systems use IoT sensors to monitor certain strengths and possible weaknesses in the building's infrastructure. These sensors can detect issues that the human eye cannot. The data collected is used to accurately predict structural issues before they occur and perform maintenance for structural stability.

- **AI and ML**: AI and ML use collected data to develop schedules and construction plans. This data is also used to predict construction projects' future outcomes, estimated completion dates, and total costs. AI systems monitor project progress, help locate assets, and help determine any onsite errors in plumbing, electrical, and mechanical systems.

Construction projects are facing productivity issues with supply chain shortages and a lack of skilled workers. The use of IoT smart construction technologies can help to fill the gaps and reduce schedules by nearly 20 percent. When implemented properly, these smart solutions can reduce errors and injuries while improving efficiency and productivity. Connected job sites loaded with IoT sensors make information regarding nearly every aspect of the operations available to everyone onsite or elsewhere.

Once the construction project is completed, we move on to the operational phase. *Chapter 2, Smart Building Operations and Controls*, explores in detail the operations and maintenance systems that are positively impacted by IoT solutions, smart building technologies, and intelligent learning capabilities.

Intelligent learning

Throughout this book, we have referenced AI, ML, and deep learning and their impact on smart buildings. In this section, we will discuss each of these in more detail and their expected contribution from smart buildings to unified buildings.

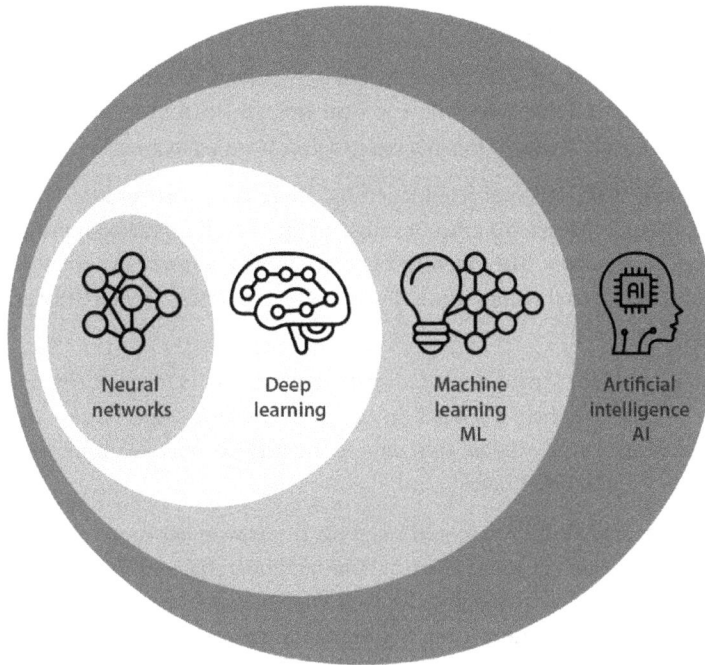

Figure 15.5 – Intelligent learning

- **AI**: AI is comprised of neural networks, deep learning, and ML, which process data to make predictions and decisions. AI can be loosely defined as machines thinking like humans.

- **Artificial neural networks** (**ANNs**): An ANN, more commonly referred to as a *neural network*, is modeled after the brain's biological neurons. Here, the neurons are called nodes, and these nodes are clustered together in layers and operate in parallel. When a node receives a computerized numerical signal, it will process it, and then it signals the other neurons connected to it. This neural reinforcement creates improved recognition of patterns and expertise and enhances overall learning.

- **Deep learning**: Deep learning is a form of ML that includes many neural network layers and massive, complex disparate data. Systems will engage with multiple layers within the network to develop increasingly higher-level outputs. Examples include pharmaceutical analysis, speech recognition, and image classification.

- **ML**: ML in the context of smart buildings refers to contributing improvements to every system and facet of the building. ML is a subset of AI whereby computers are taught to learn from data and experience instead of being programmed. Algorithms find data patterns and correlations to make predictions and decisions based on the analysis. They improve with use as more data and experience make them more accurate.

There are four models of ML:

- **Supervised**: Supervised ML uses examples and answer keys to teach the machine the expected outcome. Having seen the same outcome over time, the machine remembers that it is the expected outcome. It is taught the answer to select from an answer key.

- **Unsupervised**: With unsupervised learning, there is no answer key. Like how humans observe the world, the machine studies unstructured and unlabeled input data to identify patterns and correlations. The machine becomes more and more experienced as more data is made available. Facial recognition, market research, and cybersecurity are examples of unsupervised learning.

- **Semi-supervised learning**: Labeled and structured data is input with unstructured and unlabeled input data to help the machine to learn. The semi-supervised learning algorithm tells the machine to study the labeled data to identify correlative properties that it should then apply to the unlabeled data.

- **Reinforced learning**: With reinforced learning, the answer key used in supervised learning is replaced with a set of allowable rules, actions, and potential end states. The machine learns by experience and rewards, whereby the reward here is numerical and becomes programmed into the algorithm for the next time. Reinforced learning is like teaching someone how to play chess. It would be extremely difficult to explain every potential move; therefore, chess is taught by explaining rules, and skills are built through practice and experience.

ML algorithms can recognize correlations and patterns and are therefore very good at analyzing NOI. This allows for the immediate assessment of the smart building solution's operational impact. Predictive maintenance is another good example of how ML is used.

AI is making buildings smarter

IoT sensors help the building's components to communicate with each other, but they typically do not provide any intelligence. Without intelligence, there is just a deluge of raw data that needs to be sorted to provide operational insights. As more and more IoT sensors are implemented in buildings, more and more data will be collected that will need to be analyzed to streamline, optimize, and innovate the building's systems. AI is used to turn all this data into actionable intelligence.

AI may be used throughout the smart building and with all its systems; however, it will most often be used for the following:

- **Security**: AI is used to detect cybersecurity attacks and assist surveillance cameras in processing data to spot unusual behavior patterns.

- **Energy optimization**: Traditionally, energy optimization was performed by using after-the-fact data to analyze energy use patterns and then implementing changes to use less energy the next time. AI and predictive analytics use real-time data to deliver more proactive solutions, such as keeping building temperatures within a tighter range. This prevents the constant on-and-off cycling of the HVAC system where energy spikes every time the system is turned on.

- **Preventative maintenance**: AI is used for the day-to-day building operations along with predictive maintenance, as presented in *Chapter 2, Smart Building Operations and Controls*. It is also used for fault detection to identify anomalies and inconsistencies and then to provide actionable insights.

- **Improving occupant comfort**: By preventing faults and optimizing the building's operations using AI, occupants will experience a more reliable and comfortable environment. Wearables and mobile applications are coupled with AI to improve the way occupants interact with the building.

The use of AI in buildings is in its early adoption stage, and new technologies that embrace AI-based solutions are being introduced with regularity now.

Unified buildings

What's next for the smart building evolution? In *Chapter 1, An Introduction to IoT and Smart Buildings*, we defined a smart building as *a building that uses an integrated set of technology, systems, and infrastructure to optimize building performance and occupant experience*. We have spent numerous chapters reviewing how IoT sensors and devices, connected with computing processing capabilities and smart applications, monitor, measure, manage, and control nearly every system in a building to improve efficiency, reduce the impact on the environment, lower operation costs, increase revenue, and, most importantly, enhance the occupant's quality of experience.

Buildings have evolved from local single control systems in the early 1980s to centralized controls in the late 1990s. In the 2000s, *intelligent buildings* added some communication, databases, and analytics capabilities to allow distributed systems to communicate back to a central system with a heavy focus on green buildings, energy management, and the development of communication protocols.

Over the last decade, intelligent buildings connected powerful cloud-based software, ML, and AI applications to transition to the *smart buildings* of today, which can conserve energy, optimize designs, and predict required actions and possible failures before they occur.

While these smart building IoT sensors and applications vastly improve the overall performance and management of the building, these systems are often still proprietary, siloed, and disconnected, which leads to multiple applications being required, data that cannot communicate with other data, and ultimately, user frustration. These smart building solutions are solving one issue at a time, while collectively creating another greater problem, a need to have a single platform to monitor, manage, and control the growing number of IoT sensors, devices, and systems.

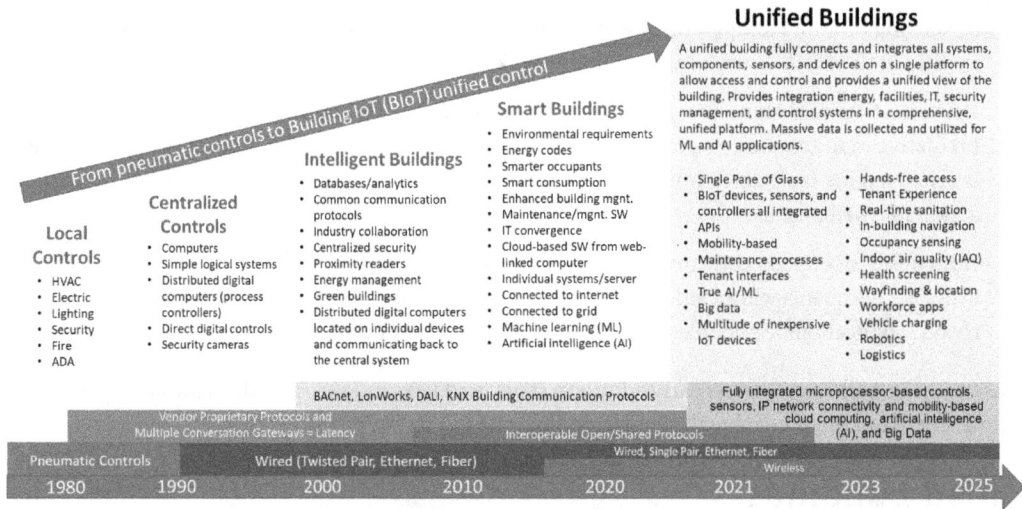

Unified Buildings

A unified building fully connects and integrates all systems, components, sensors, and devices on a single platform to allow access and control and provides a unified view of the building. Provides integration energy, facilities, IT, security management, and control systems in a comprehensive, unified platform. Massive data is collected and utilized for ML and AI applications.

From pneumatic controls to Building IoT (BIoT) unified control

Smart Buildings
- Environmental requirements
- Energy codes
- Smarter occupants
- Smart consumption
- Enhanced building mgnt.
- Maintenance/mgnt. SW
- IT convergence
- Cloud-based SW from web-linked computer
- Individual systems/server
- Connected to internet
- Connected to grid
- Machine learning (ML)
- Artificial intelligence (AI)

Intelligent Buildings
- Databases/analytics
- Common communication protocols
- Industry collaboration
- Centralized security
- Proximity readers
- Energy management
- Green buildings
- Distributed digital computers located on individual devices and communicating back to the central system

Centralized Controls
- Computers
- Simple logical systems
- Distributed digital computers (process controllers)
- Direct digital controls
- Security cameras

Local Controls
- HVAC
- Electric
- Lighting
- Security
- Fire
- ADA

- Single Pane of Glass
- BIoT devices, sensors, and controllers all integrated
- APIs
- Mobility-based
- Maintenance processes
- Tenant interfaces
- True AI/ML
- Big data
- Multitude of inexpensive IoT devices

- Hands-free access
- Tenant Experience
- Real-time sanitation
- In-building navigation
- Occupancy sensing
- Indoor air quality (IAQ)
- Health screening
- Wayfinding & location
- Workforce apps
- Vehicle charging
- Robotics
- Logistics

BACnet, LonWorks, DALI, KNX Building Communication Protocols

Fully integrated microprocessor-based controls, sensors, IP network connectivity, and mobility-based cloud computing, artificial intelligence (AI), and Big Data

Vendor Proprietary Protocols and Multiple Conversation Gateways = Latency

Interoperable Open/Shared Protocols

Pneumatic Controls | Wired (Twisted Pair, Ethernet, Fiber) | Wired, Single Pair, Ethernet, Fiber / Wireless

1980 | 1990 | 2000 | 2010 | 2020 | 2021 | 2023 | 2025

Figure 15.6 – Evolution to unified buildings

Today, each smart building solution comes with its own interface, via either a smartphone or a website application, which forces building operators to jump between hundreds of different management systems. Smart building management and automation systems need to evolve to a fully integrated and centralized platform to manage the building's control systems, while providing information to all stakeholders to create a truly unified building.

I imagine that the next evolution of smart buildings will be the *unified building*. A unified building will fully connect and integrate all systems, components, sensors, and devices on a single platform to allow access and control and provide a unified view of the building on a **single pane of glass** (**SPOG**). This will provide for the full integration of energy, facilities, IT, security management, and control systems in a comprehensive, unified platform. Fully integrated and connected microprocessor-based controls and sensors will deliver massive amounts of data utilized for ML and AI applications.

Smart buildings of the future will be fully connected, dynamic, centralized, and agile ecosystems utilizing massive amounts of data and powerful automation. The evolution to a unified building will result in delivering common features such as the following:

- **Fully integrated**: A fully integrated, microprocessor-based, sensor-, control-, IP network connectivity-, and mobility-based cloud computing system supporting big data and AI systems.

- **Full asset visibility in real time**: A single management system that monitors, tracks, controls, and reports on all the building's connected assets and provides a granular view of full site operations. Multi-site operations and portfolio management deliver high-performance asset management capabilities.

- **SPOG**: A single user interface to deliver full asset visibility in a single holistic view with data displayed in understandable and actionable ways. It is the beginning point where operators can get a sense of the greater picture of what is occurring in their building. SPOG was covered in detail in *Chapter 6, The Smart Building Ecosystem*.

- **Unified security**: A unified building enables continuous monitoring and management of the building along with full security management interfaces, which allow for the monitoring of both physical security and cybersecurity throughout the building and the network. The unified building will provide the ability to detect anomalies or deviations that require an immediate response and take necessary action to mitigate or block them. IoT sensors will collect real-time data to enable predictions and reactions.

- **Cross-domain innovation**: New, innovative solutions are created and delivered when traditionally segmented activities are brought together. Horizontal expertise is shared, fostering innovation in the IoT sensor, network, management, and security levels. For example, energy, facility, IT, control systems, and security management systems can be brought together to collectively predict energy consumption requirements.

- **Holistic KPIs**: KPIs can combine correlated building operations and security events. These KPIs could be established from four primary domains: energy supply and efficiency, information and communication, building automation, and safety and security. The holistic KPIs permit security and operations managers to have unified access to optimize the building's energy supply, monitor physical security and cybersecurity, and monitor the IT infrastructure simultaneously.

- **Adaptive demand and response**: Effective and secure resource forecasting and management is what operators rely on and is the *smartness* to conduct holistic building monitoring, management, and optimization. Combining real-time energy management and day-to-day operations to achieve targets may mean striking a balance between energy consumption and occupant comfort levels.

- **Contextual HMI**: With the integration of IoT sensors connected via cloud technology, different sensors use different portals for information access. This makes managing the operations difficult without the full context and reduces situational awareness. A contextual **human-machine interface** (HMI) allows building operators to have KPIs and crucial real-time information at their fingertips in the context of their job and at their current physical location.

- **Context-specific management**: Buildings are becoming more multi-purposeful with mixed-use events and co-existence. A recent trend shows the upper floors of commercial buildings being converted to residential units, thus creating a mixed-use building with systems originally designed and engineered for one purpose. This creates variability in energy consumption and security requirements that require context-specific management.

Buildings designed for temporary events such as concerts, sports, and cultural events also require context-specific management. Each will need to account for sources of energy, adaptive consumption patterns, demand response, load shedding, and direct load control to address the changing operational conditions.

The unified building will deliver people-centric outcomes to enhance occupant safety and well-being. Unifying IoT sensors around lighting, indoor air quality, ventilation, and temperature can result in reduced absenteeism, improved retention, and better staff motivation. Unifying systems and data can positively impact workflows and supply chains, and create an agile workplace that is more energy efficient and produces less waste.

IoT will drive the future of buildings

The **Pacific Northwest National Laboratory** (**PNNL**) has long been considered the leading authority studying and defining the impact of technology, energy, and data analytics on buildings. They worked very closely with their partners, academia, industry leaders, and the US Department of Energy to develop a 2022 report titled *A Vision for Future Buildings*, which begins to envision what buildings will be like in 100 years.

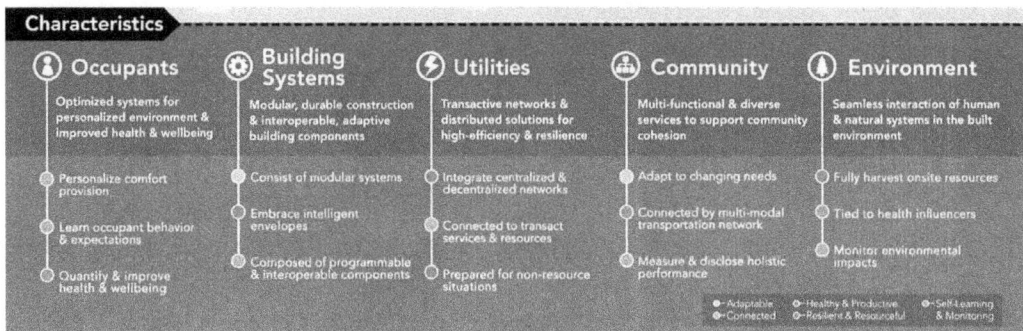

Figure 15.7 – PNNL vision characteristics (image courtesy of PNNL)

This integrated and coherent vision identified five characteristics and targets for achieving future building goals. They are as follows:

- **Occupants**: Systems are optimized for occupant expectations and behavior to deliver personalized environments and improve occupants' health and well-being. Biometric data captured with IoT sensors will learn occupant preferences to automatically make adjustments to an individual's environment. The goal is to maximize personal comfort, productivity, and health while minimizing wasted resources. Wearable and portable devices are integrated with building systems to personalize occupants' environments relating to lighting, temperature, and comfort settings.

- **Building systems**: Buildings will need to be made easily updatable by using modular, interoperable, durable, and universally compatible components. Buildings will become reconfigurable whereby the same building could be used as a meeting space one day and as a residential unit or restaurant the next. Buildings will become fluid with plug-and-play and programmable systems. The materials used for building construction will be able to generate and capture solar energy, collect rainwater, control light levels, regulate temperature, and filter the air to create the smart building envelope.

- **Utility**: Future building technology will literally and virtually connect buildings and utility systems in a transactive network with distributed solutions. This integrated and distributed utility network will capture energy from a variety of different sustainable energy sources, such as the sun, wind, rain, plants, and energy produced from other buildings. Buildings are connected to trade or share resources and building services such as water purification, air-cleaning, and onsite waste management.

- **Community**: The transfer of water, waste, products, and services will be streamlined. Buildings will be connected via a multi-model transportation network to free up resources that are currently wasted to meet individual building needs. Building performance will be measured and reported holistically.

- **Environmental**: Environmental impacts will be measured and monitored in real time and building owners will be able to accurately determine the cost of ownership. Onsite resources will be fully harvested and the interaction between humans and natural systems will be seamless.

Buildings are large, long-term investments that have a major impact on the environment. Additionally, they are where people spend more than 90 percent of their time and so it makes sense to explore the functions and capabilities of what future buildings could be. These characteristics can be used as the framework for public and private leaders to advance research and development today to create the buildings of the future.

What's next

New IoT-based devices are continually being developed and we should recognize that almost any physical item can now be made smart by adding IoT sensors, some connectivity, and computing capabilities. If the return-on-investment model is favorable, chances are it will become smart. IoT device costs will continue to decrease, making more and more items candidates for being made smart.

The following are a few examples of what is trending at the time of this writing:

- **Matter**: 2023 will see the introduction of Matter version 1.0 from the **Connectivity Standards Alliance (CSA)**, formerly the Zigbee Alliance. This 280-member alliance includes giants such as Google Assistant, Apple HomeKit, and Amazon Alexa. They developed a smart home standard that allows their applications to communicate with each other and increase their cybersecurity capabilities.

Figure 15.8 – CSA Matter standard

- **Smart plug-and-play power jacks**: We will soon see standardized plug-and-play power jacks (such as **SkyX Plug**) that are installed in ceilings to support plugged-in IoT devices weighing up to 200 pounds. These devices range from Bluetooth speakers, smoke/CO_2 detectors, heating and cooling devices, Wi-Fi extenders, cameras, and any other IoT device that needs to be mounted. Electrical conversion kits will be used to retrofit existing ceiling electrical boxes. Soon, lighting and ceiling fan manufacturers will develop adaptor kits and other manufacturers will develop new SkyX Plug-equipped gear.

- **Tiny smart sensors**: Tiny sensors that can be attached to any physical object and connected wirelessly are becoming commonplace in smart buildings. These tiny form factor sensors are disrupting industry because they can be cost-effectively installed in every nook and cranny of a building and monitor and measure anything. The extended battery life for these tiny sensors is 13.1 to 15 years. End-to-end encryption allows for open APIs and enhanced security.

- **3D printing**: 3D printing in the IoT space shows trends of significantly reduced costs and improved supply chain support. 3D printing is leading to a wide variety of new IoT devices being developed at a much faster pace than traditional manufacturing. Manufacturing incredibly tiny and delicate parts, such as the tiny IoT sensors mentioned earlier, is much easier to do with 3D printers. This makes IoT more affordable and accessible to speed up the implementation in buildings.

- **Smart building facades**: Building facade materials are being embedded with sensing and condition regulation by companies such as Dutch start-up *PHYSEE's SmartSkin* solution. Solar cells and weather sensors are used along with their patented nano-grid technology to convert solar energy into electricity. They use the building's envelopes to produce sustainable energy while managing the occupant's comfort levels.

- **Smart flooring analytics**: Smart floors use sensors to gain insights into people's walking patterns. Retail, medical, residential, and industrial applications capture data to determine favored routes, occupancy, abandonment, waiting time, and traffic flow. Low-energy smart mats are daisy-chained to create a smart floor.

- **Smart ventilation systems**: Ventilation systems use IoT sensors to regulate airflow parameters to reduce a building's heating or cooling demand. These sensors also aid with predictive maintenance capabilities to detect faults in advance. **Passive Ventilation with Heat Recovery (PVHR™)** by UK-based Ventive is a patented method to deliver fresh air while reducing heat loss by using a natural ventilation system. This is completed by transforming thermal heat from exhaust air into fresh incoming air.

- **Smart green roofing**: Natural vegetation from trees and gardens planted on building rooftops assists in energy conservation, runoff management, as well as providing roofing reinforcement and sequestration of carbon. IoT plant sensors, actuators, weather data, simulation models, and predictive analytics deliver real-time vegetation monitoring and management.

- **AI-driven digital twins**: In *Chapter 8*, *Digital Twins – a Virtual Representation*, we discussed how building-related documentation and drawings are being digitized to store, maintain, and make them readily available. Digital twins create replicas of buildings and building systems. Combining AI and ML with digital twin technology enables real-time monitoring, predictive analytics, and many other tools for building optimization. Natural language search, AI-driven cost analysis, multi-layer data operations, and 3D information are combined for smart building life cycle management.

- **Big data**: Big data is a by-product of building smart IoT networks in buildings. The collection of massive amounts of data from IoT technologies requires data science and AI processes to capture, curate, manage, and analyze data to obtain valuable insights and deliver real-time outcomes. It will play a substantial role in the feasibility of smart building initiatives to improve building operations.

These examples are just a small representation of what's to come to advance smart buildings. None of these are possible without the use of IoT sensors and devices. In 2022, the global smart buildings market was valued at USD 82.69 billion and is expected to reach 143.373 billion in 2028 with a CAGR of 9.61 percent according to Market Watch.

This explosive growth is driven by the surge in connected IoT devices with an estimated 11.3 billion IoT devices in 2021 expected to grow to 29.4 billion by 2030. AI and ML will be critical to building operations, and cybersecurity will be the biggest priority.

Final thoughts – data tagging

New IoT sensors and devices will continue to evolve and almost every aspect of a building will become smart, generating massive amounts of data. My biggest concern going forward centers around data tagging and the current lack of a disciplined tagging convention within the industry. Like the issues created by proprietary disparate building systems of the last generation, engineers are creating their own naming and data tagging conventions to describe various elements within a building.

Smart buildings need a data tagging standard that uniquely defines every element within a building to create a common language. Data tagging standards will allow data flow to move quicker and more efficiently between devices, controllers, and interrelated equipment. BrickSchema and Project Haystack were developed to handle these issues; however, they each need to be applied manually, resulting in a very time-consuming and costly process that is prone to errors.

The **Ontology Alignment Project** (**OAP**) is one attempt to standardize tagging through an open sourced taxonomy for the built environment that automates the process. Data tagging and naming standards facilitate integration that normalizes and standardizes data from IoT devices. OAP enhances supervisory controls, analytics software, diagnostics, fault detection, and other functions that use building data. It is also an API that allows users to build software applications to mesh with the OAP data model. OAP can be found on GitHub, and the open source team is working closely to align with Project **Haystack 4** (**H4**). The industry must align on an ontology for modeled data.

Summary

In this chapter, we reviewed how NOI is used to measure and report the return on smart building initiatives. While all the solutions discussed throughout this book can be applied to nearly every building type, we specifically explored how smart buildings are digitalizing the commercial real estate market. Advances in smart building construction materials, processes, tools, and equipment are helping to improve safety, quality, and lower costs while reducing the time required to build buildings.

AI, ML, and deep learning will play an even bigger role in the evolution of smart buildings to unified buildings. The increased number of cheaper IoT sensors will generate massive amounts of data that needs to be processed using intelligent systems. We explored how IoT will drive the future of buildings and reviewed a framework developed to achieve the long-term vision for a building. Lastly, we listed a few examples of what we believe the next IoT-driven solutions will be and their impact on the smart building industry.

We have made it to the end, and I thank you for your commitment to finishing. I'm sure that in the time it took to write this book, some of the IoT and smart building technologies have evolved to the next level already, as everything is moving quickly. I hope that you found this book comprehensive and informative, but most of all, I hope that you can implement some of the IoT and smart building ideas immediately. The road to a smart building is a journey that begins with providing IoT and smart building technology education to achieve support from your stakeholders. I encourage you to have key stakeholders obtain a copy of this book to help support your efforts.

Index

T

‹packt›

Packtpub.com

Subscribe to our online digital library for full access to over 7,000 books and videos, as well as industry leading tools to help you plan your personal development and advance your career. For more information, please visit our website.

Why subscribe?

- Spend less time learning and more time coding with practical eBooks and Videos from over 4,000 industry professionals

- Improve your learning with Skill Plans built especially for you

- Get a free eBook or video every month

- Fully searchable for easy access to vital information

- Copy and paste, print, and bookmark content

Did you know that Packt offers eBook versions of every book published, with PDF and ePub files available? You can upgrade to the eBook version at packtpub.com and as a print book customer, you are entitled to a discount on the eBook copy. Get in touch with us at customercare@packtpub.com for more details.

At www.packtpub.com, you can also read a collection of free technical articles, sign up for a range of free newsletters, and receive exclusive discounts and offers on Packt books and eBooks.

Other Books You May Enjoy

If you enjoyed this book, you may be interested in these other books by Packt:

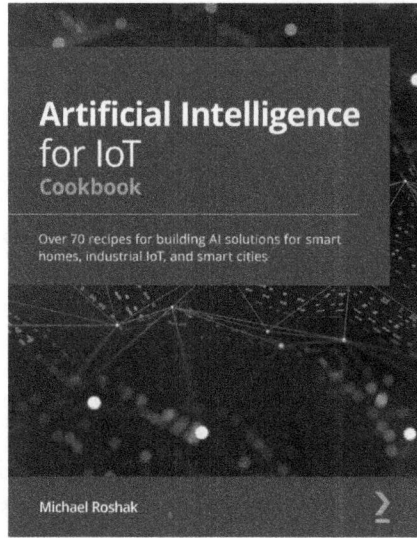

Artificial Intelligence for IoT Cookbook

Michael Roshak

ISBN: 978-1-83898-198-3

- Explore various AI techniques to build smart IoT solutions from scratch
- Use machine learning and deep learning techniques to build smart voice recognition and facial detection systems
- Gain insights into IoT data using algorithms and implement them in projects
- Perform anomaly detection for time series data and other types of IoT data
- Implement embedded systems learning techniques for machine learning on small devices
- Apply pre-trained machine learning models to an edge device
- Deploy machine learning models to web apps and mobile using TensorFlow.js and Java

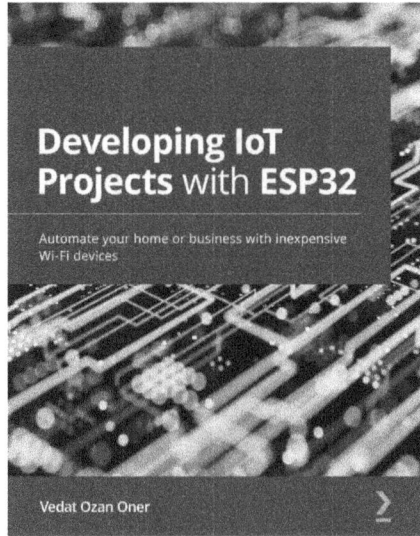

Developing IoT Projects with ESP32

Vedat Ozan Oner

ISBN: 978-1-83864-116-0

- Explore advanced use cases like UART communication, sound and camera features, low-energy scenarios, and scheduling with an RTOS
- Add different types of displays in your projects where immediate output to users is required
- Connect to Wi-Fi and Bluetooth for local network communication
- Connect cloud platforms through different IoT messaging protocols
- Integrate ESP32 with third-party services such as voice assistants and IFTTT
- Discover best practices for implementing IoT security features in a production-grade solution

Packt is searching for authors like you

If you're interested in becoming an author for Packt, please visit `authors.packtpub.com` and apply today. We have worked with thousands of developers and tech professionals, just like you, to help them share their insight with the global tech community. You can make a general application, apply for a specific hot topic that we are recruiting an author for, or submit your own idea.

Share Your Thoughts

Now you've finished *Internet of Things for Smart Buildings*, we'd love to hear your thoughts! Scan the QR code below to go straight to the Amazon review page for this book and share your feedback or leave a review on the site that you purchased it from.

https://packt.link/r/1804619868

Your review is important to us and the tech community and will help us make sure we're delivering excellent quality content.

Download a free PDF copy of this book

Thanks for purchasing this book!

Do you like to read on the go but are unable to carry your print books everywhere?

Is your eBook purchase not compatible with the device of your choice?

Don't worry, now with every Packt book you get a DRM-free PDF version of that book at no cost.

Read anywhere, any place, on any device. Search, copy, and paste code from your favorite technical books directly into your application.

The perks don't stop there, you can get exclusive access to discounts, newsletters, and great free content in your inbox daily

Follow these simple steps to get the benefits:

1. Scan the QR code or visit the link below

https://packt.link/free-ebook/9781804619865

2. Submit your proof of purchase
3. That's it! We'll send your free PDF and other benefits to your email directly

www.ingramcontent.com/pod-product-compliance
Lightning Source LLC
Chambersburg PA
CBHW080515220326